Debug Automation from Pre-Silicon to Post-Silicon

Mehdi Dehbashi • Görschwin Fey

Debug Automation from Pre-Silicon to Post-Silicon

 Springer

Mehdi Dehbashi
Institute of Computer Science
University of Bremen
Bremen, Germany

Görschwin Fey
Institute of Space Systems
German Aerospace Center
Bremen, Germany

ISBN 978-3-319-09308-6   ISBN 978-3-319-09309-3 (eBook)
DOI 10.1007/978-3-319-09309-3
Springer Cham Heidelberg New York Dordrecht London

Library of Congress Control Number: 2014948782

Printed on acid-free paper

Springer is part of Springer Science+Business Media (www.springer.com)

Preface

The application of Very Large Scale Integration (VLSI) circuits is ubiquitous while the size of the hardware components is shrinking. VLSI circuits are used for a wide range of different applications in embedded systems such as medical electronics, automotive systems and avionics. A failure of a chip in non-critical applications may cause significant economic loss while in critical applications may also endanger human's lives in the worst case. Consequently, the correct design of VLSI circuits is crucial.

The debugging process is dedicated to localize and to rectify the root cause of the erroneous behavior of VLSI circuits. This process often remains as a manual task and increases the time required for development cycles of Integrated Circuits (ICs) significantly. Therefore, debug automation procedures are required to accelerate finding and fixing bugs and faults and, consequently, to increase the productivity of IC design.

This book contributes to debugging and diagnosis technology at the most challenging gaps on different abstraction levels of a hardware system, i.e., chip, gate-level, Register Transfer Level (RTL) and transaction-level.

In this book, we propose automated debugging approaches for the bugs and the faults which appear in different abstraction levels of a hardware system, i.e., transaction-level, RTL, and gate-level. Our automated debug approaches are applied to a hardware system at different granularities to find the possible location of bugs and faults. The transaction-based debug approach is applied to a hardware system at transaction-level asserting the correct relation of transactions. Our automated debug approach for design bugs finds the potential fault candidates at RTL and gate-level of a circuit. In this case, logic bugs and synchronization bugs are considered as they are the most difficult bugs to be localized. For electrical faults and, in particular, delay faults our proposed debug automation finds the potential failing speedpaths in a circuit at gate-level.

The proposed debug approaches for transactions, design bugs and electrical faults have been evaluated on suitable benchmarks at different levels of abstraction, i.e., transaction-level, RTL and gate-level. The experiments have shown that our debug approaches achieve high diagnosis accuracy and reduce the debugging time. As a result, the time of the IC development cycle decreases and the productivity of IC design increases.

We would like to thank all our coauthors for the fruitful collaboration. In particular, we would like to thank Dr. André Sülflow for the constructive discussions and his support during our work on this book and the underlying techniques. We are grateful to Prof. Kaushik Roy and Prof. Anand Raghunathan for their collaboration, important discussions, and comments. We would like to thank the German Research Foundation (DFG) for funding our work.

Munich, Germany   Mehdi Dehbashi
Bremen, Germany   Görschwin Fey
June 2014

Contents

Acronyms

ACE	Access-Control Extension
AD	Automated Debugging
BDD	Binary Decision Diagram
BMC	Bounded Model Checking
CDU	Central Debug Unit
CNF	Conjunctive Normal Form
CUD	Core-Under-Debug
cv	Controlling value
DFD	Design-For-Debug
DFT	Design-For-Test
DRI	Debug Redundant Information
DTG	Diagnostic Trace Generation
DTV	Diagnostic Trace Validation
DU	Debug Unit
EoRp	End of Response
EoRq	End of Request
ErrRp	Response Error
ErrRq	Request Error
FSM	Finite State Machine
HDL	Hardware Description Language
IC	Integrated Circuit
LBA	Local Branch Activation
LDU	Local Debug Unit
LMBA	Limited Minimization followed by Branch Activation
MaxSAT	Maximum SATisfiability
MBD	Model-Based Diagnosis
MSPI	Minimization of Sensitized Path Intersection
NI	Network Interface
ncv	Non-controlling value

NoC	Network-on-Chip
PI	Primary Input
PO	Primary Output
PT	Propagation Time
QBF	Quantified Boolean Formula
RTL	Register Transfer Level
SAT	Boolean Satisfiability
SBST	Software-Based Self-Testing
SoC	System-on-Chip
SoRp	Start of Response
SoRq	Start of Request
SSTA	Statistical Static Timing Analysis
STA	Static Timing Analysis
TAM	Time Accurate Model
TC	Time Control
TDPSL	Transaction Debug Pattern Specification Language
VC	Variation Control
VL	Variation Logic

Symbols

AT_{max}	Maximum arrival time
AT_{min}	Minimum arrival time
C	Circuit graph
CE	Counterexample
CEs	Set of counterexample
D	Maximum timing variation
D_l	Longest path delay
D_s	Shortest path delay
DT	Diagnostic trace
DTs	Set of diagnostic traces
E	Set of edges
ET	Erroneous trace
$\mathscr{F}$	Set of fault candidates
FC	Fault candidate
g	Gate $\in V$
I	Set of primary inputs
$i_1, \ldots, i_n$	Primary inputs of C
O	Set of primary outputs
$o_1, \ldots, o_m$	Primary outputs of C
P	Structural path of C
S	Set of states
SEL	Set of select lines
sel	Select line of multiplexer
T	Clock period
V	Set of circuit nodes
v	(Boolean) value
X	Don't care value
$x_1, \ldots, x_n$	Boolean variables
ω	Clause of Φ

Φ	CNF, set of clauses
$\cdot$	Boolean AND operator
$+$	Boolean OR operator
$\overline{}$	Boolean NOT of $\cdot$
$;$	Concatenation operator

Chapter 1
Introduction

The presence of *Very Large Scale Integration* (VLSI) circuits in our daily life increases while the size of the hardware components is shrinking. VLSI circuits are used for different applications in embedded systems such as medical electronics, automotive systems and avionics. A failure of a chip in non-critical applications may cause significant economical loss while in critical applications may also threaten the human life in the worst case. Consequently, the correct design of VLSI circuits is crucial.

The VLSI system design methodology starts with a system design team writing the specification of the system as a text. Then, the system model is implemented and the concepts as well as the algorithms at the system level are verified [Ver02]. The common languages to describe the system model of hardware are C, C++, and SystemC. Figure 1.1 shows the simplified design flow in which the first step is the system model implementation.

After validating the specification of an *Integrated Circuit* (IC), the functions of the specification are implemented by hardware description languages such as Verilog and VHDL (second step in Fig. 1.1). The hardware functions are implemented at *Register Transfer Level* (RTL) using hardware description languages. RTL is a level of design abstraction modeling the flow of digital signals (data) between hardware registers and the logical operations performed on these signals.

Having an RTL description (design) and a specification of the system, verification tools check the functional correctness of the design against the specification. The verification proves that the design is consistent with the available specification. This process is performed using simulation-based or formal approaches. When an inconsistency between the specification and the design is detected by the verification tool, this inconsistency or erroneous behavior is returned as a *counterexample*. The counterexample shows the erroneous behavior of the design with respect to the specification.

Having a counterexample, debugging starts to find potential locations of a design bug. A mistake or a problem in a hardware design is called *bug*. Debugging at the

© Springer International Publishing Switzerland 2015
M. Dehbashi, G. Fey, *Debug Automation from Pre-Silicon to Post-Silicon*,
DOI 10.1007/978–3–319–09309-3_1

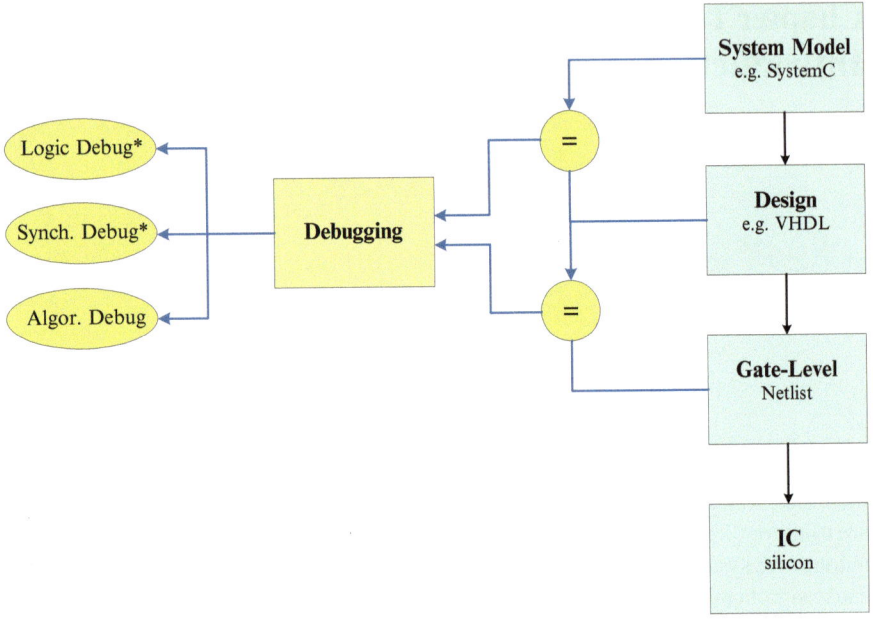

Fig. 1.1 Abstraction levels in the design flow and pre-silicon debugging

design level before fabricating a chip on the silicon is called *pre-silicon debugging*. The left part of Fig. 1.1 shows the steps of the pre-silicon debugging. The debugging process is dedicated to localize and to rectify the root cause of the erroneous behavior. This process often remains as a manual task and increases the time of the IC development cycle significantly. A deviation of the design behavior from the specification behavior is called an *error*. Automated debugging identifies the potential sources of the observed errors shown by counterexamples. The output of automated debugging is a set of *fault candidates*. Each fault candidate is a set of components of the design which may be modified to fix the erroneous behavior of the counterexamples. There are different approaches for debug automation. Automated approaches for pre-silicon debugging rely on simulation [VH99], *Binary Decision Diagrams* (BDD) [CH97], and *Boolean Satisfiability* (SAT) [SVAV05]. Among these approaches, debugging based on SAT has been shown as a robust and effective approach in a variety of design scenarios from diagnosis to debugging properties. However, SAT-based debugging needs certain improvements to gain higher diagnosis accuracy.

Design bugs at RTL are divided into three main classes: logic bugs, algorithmic bugs, and synchronization bugs [CMA08]. In the following, these three classes of design bugs are described:

- Logic bugs—Erroneous logic in combinational circuits characterizes the class of logic bugs. A logic bug may occur because of a mistake which a designer may make while writing the hardware code. A designer may write an erroneous logic block. For example, he uses an AND gate instead of an OR gate, or a NOT gate instead of a BUF gate. Also an extra wire may be added to a logic block, or there may be a missing wire in a logic block. This class includes bugs in the logic of the combinational circuit. On RTL logic bugs are similarly likely, e.g., as erroneous conditions.

* Algorithmic bugs—This class of design bugs belongs to the algorithmic implementation of a design. These bugs occur when a designer implements the algorithm of a hardware module in a wrong way. Algorithmic bugs usually require major modifications to be fixed. To fix the incorrect implementation of the algorithm of a module, a new module should be implemented. Fixes for algorithmic bugs are not limited to the modification of the combinational circuit. They require sometimes multiple major modifications changing the whole module [CMA08]. This class includes bugs which influence both the combinational circuit and the sequential circuit of a module.

* Synchronization bugs—An error in the timing behavior of the implementation belongs to the class of synchronization bugs. A synchronization bug is a case where a signal needs to be latched some cycles earlier or some cycles later in order to keep the correct timing of signals in the implementation of a design.

After verifying the design against the specification and fixing the bugs in the design, logic synthesis is performed. Logic synthesis converts a design to a gate-level circuit. There are some automated tools for logic synthesis and thus this process is usually automated. When a synthesis tool converts a design to a gate-level circuit, the new circuit is checked against the design. Comparing the behavior of the design and the circuit is called equivalence checking. An equivalence checker verifies the behavior of a circuit in different abstraction levels. When an inconsistency is observed between the design and the circuit at the gate-level, the corresponding inconsistency is used by debugging to find the potential fault candidates in the gate-level description of the circuit.

The transistor level design is created by a place-and-route process for chip manufacturing. Then the design is fabricated in silicon as a chip. The process to validate and to debug a fabricated chip is called *post-silicon validation and debugging* (Fig. 1.2). The post-silicon validation process is started by applying test vectors to the IC or by running a test program, such as end-user applications or functional tests, on the IC until an error is detected [CMB07a, PHM09]. The erroneous behavior and the golden responses obtained by system simulation are utilized by post-silicon debugging to find the root cause of the observed error. However, post-silicon debugging is usually manual and needs a large effort. Therefore, debug automation procedures are also required at this level to accelerate finding and fixing the bug and consequently to increase the productivity of IC products.

A challenge in the post-silicon stage is distinguishing electrical faults and design bugs. A *fault* indicates a physical flaw or defect in a chip. Design bugs may appear in the final IC product. Due to the fact that the cost of VLSI systems

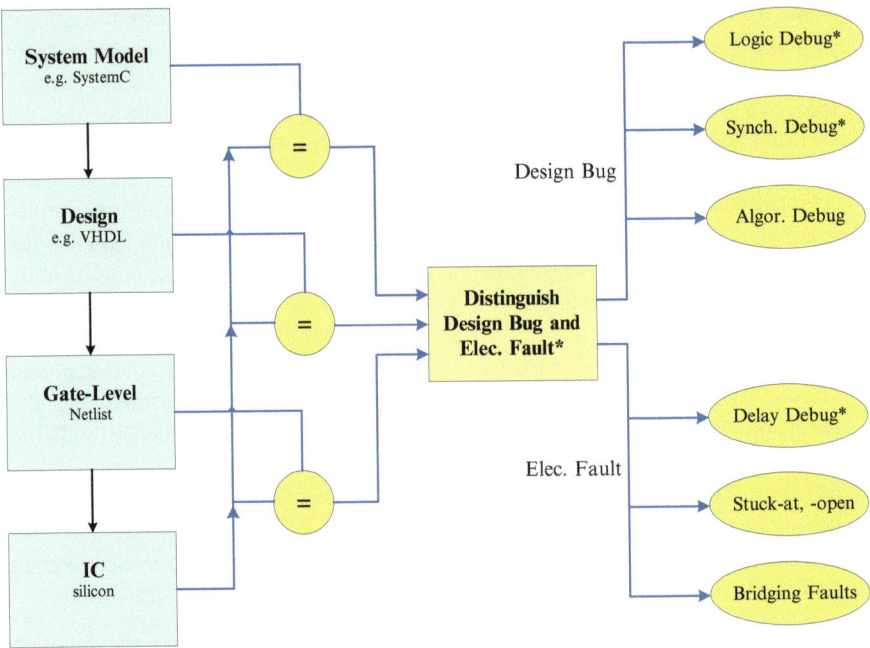

Fig. 1.2 Abstraction levels in the design flow and post-silicon debugging

verification and debugging has significantly increased in the recent years as design
size and complexity have increased. Also due to time-to-market constraints, 100 %
verification coverage at the design level is an elusive task. In this case, automated
debugging approaches identifying different kinds of bugs and faults can effectively
help to reduce the development time of IC products.

Electrical faults are caused by fabrication defects and environmental effects
such as temperature and voltage fluctuations. Fault models describe the behavior
of electrical faults in order to efficiently investigate the effect of electrical faults
at different levels of system abstraction. The most common fault models are the
following ones which abstract precise electrical fault effects to the gate-level:

- Stuck-at/stuck-open fault model—In the single stuck-at fault model, it is assumed
  that only one line in the circuit is faulty at a time. The stuck-at fault is a permanent
  fault. A stuck-at fault ties the faulty node either to the supply voltage VCC (stuck-
  at-1) or to the ground GND (stuck-at-0). Only the lines (wires) in the circuit are
  affected by this kind of fault. The function of the gates in the circuit remains
  unaffected by the fault [Hay85, Mal87]. In the stuck-open fault model, a single
  physical line in the circuit is broken. The resulting unconnected node is not tied
  to either VCC or GND. As a consequence, the node may be have like a state
  element keeping a certain logical value depending on the surrounding nets.

- Bridging fault model [MJR86]—In this fault model, two nodes of a circuit are shorted. Typically, three classes of bridging faults are considered: (1) bridging within a logic element (transistor gates, sources, or drains shorted); (2) bridging between logic nodes (i.e. inputs or outputs of logic elements) without feedback; (3) bridging between logic nodes with feedback.
- Delay fault model [WLRI87, LS80]—In this fault model, the logic function of the circuit is error free. Some physical defects, such as process variations, may cause delays of gates in the circuit to exceed the predefined bounds. Temperature and voltage fluctuation, wear-out and aging also affect the timing behavior of the circuit. Two delay fault models are typically used: gate delay fault model and path delay fault model. The gate delay fault model considers a slowdown for each gate which logically delays either a rising or falling transition on the output of the corresponding gate in the circuit, i.e., slow-to-rise and slow-to-fall. The path delay fault model is a fault model in which the total delay along a path from input to output in a circuit exceeds the timing constraint.

Debug automation for electrical faults and especially delay faults is lacking. However, having automatic debug processes to localize delay faults can help analyzing the physical root causes of the fault and consequently improving the production process and the performance of VLSI circuits.

Post-silicon debug methods usually concentrate on functional parts of systems. But nowadays the complexity of systems has increased both in functionality of each core and also in communication among cores [GF09]. A modern *System-on-Chip* (SoC) has many processors, IP cores, and other functional units. In this case, suitable debug infrastructure is required to efficiently handle errors related to complicated communication of SoCs. A suitable debug infrastructure can help to automate debugging for complex SoCs.

This book contributes to debugging and diagnosis technology at the most challenging gaps on different abstraction levels of a hardware system, i.e., chip, gate-level, RTL and transaction level. In the following presentation we typically treat debugging and diagnosis in a unified way. Moreover, for the developer it is often unknown whether she is hunting for a design bug or trying to diagnose a physical fault. Thus, we refer to this process as debugging in most cases. The main contributions of this book are as follows:

- Proposing an automated debug approach for logic bugs at the pre-silicon stage to increase the diagnosis accuracy.
- Debug automation for post-silicon debugging of logic bugs using trace buffers, *Model-Based Diagnosis* (MBD) and diagnostic traces.
- Introducing a generalized debug automation approach which can be used at both pre-silicon and post-silicon stages.
- Proposing an automated debug approach for synchronization bugs which is able to differentiate synchronization bugs and logic/algorithmic bugs.
- Modeling timing variations in the functional domain.
- Debug automation for delay faults using timing variation models.

- Integrating *Static Timing Analysis* (STA) and functional analysis to efficiently debug delay faults.
- Developing a debug infrastructure to online debug of SoCs at the transaction-level.

Figures 1.1 and 1.2 show how our methodology is embedded into the design flow at pre-silicon and post-silicon. The sections labeled with a star in Figs. 1.1 and 1.2 highlight our contributions. As algorithmic bugs cause major changes in the behavior of a module of the design in comparison to the specification [CMA08], this kind of bugs can be easily localized. However, as logic bugs and synchronization bugs affect the small components of the circuit, the error localization is difficult for this kind of bugs. Therefore, in pre-silicon debugging, we focus on the debug automation for logic bugs and synchronization bugs. In post-silicon debugging, after fabricating a chip, we first propose an approach to distinguish design bugs and electrical faults as shown in Fig. 1.2. Then we automate debugging for both design bugs and electrical faults. For electrical faults, our focus is on delay faults which are more difficult to be localized as in this case the logic function of the circuit is error free. For stuck and bridging fault models which affect the logic function of the circuit, the fault model can be utilized by MBD to localize the faults [SVAV05]. To automate debugging for design bugs and electrical faults, we utilize SAT-based debugging as core technology. For debugging, the focus of this book is localization of faults and bugs and increasing the diagnosis accuracy. At the post-silicon stage, an on-chip infrastructure for transaction-based debug is proposed to debug design bugs and electrical faults which affect the communication in a large SoC. Our debugging approaches are validated on suitable benchmarks at transaction-level, RTL and gate-level.

In the following chapters, we first present the preliminaries on circuits and SAT. Then basic SAT-based debugging is explained. Also the concept of transactions in *Transaction Level Modeling* (TLM) and *Transaction Debug Pattern Specification Language* (TDPSL) are introduced. Then the book is divided into three parts. At the end of the book, the list of acronyms and symbols are mentioned.

The focus of Part I is automated debugging of design bugs. This part presents our contribution for debug automation of logic bugs and synchronization bugs. For logic bugs, our approach integrates verification and debugging in a unified flow to increase diagnosis accuracy. Diagnostic traces are utilized to bridge the gap between verification and debugging. Also a unified framework is proposed for both pre-silicon and post-silicon debugging of logic bugs.

Part II presents our contributions to automate debugging of delay faults. Our approaches automate debugging for delay faults at post-silicon. To achieve our goal, first we model the timing variations. Then we build a suitable correction block according to the timing variation model. Next, our automated approach based on SAT finds the fault candidates automatically.

A large SoC includes the components for both computation and communication. In Parts I and II we present automated approaches which are utilized for computational components of an SoC. In Part III we focus on the components for the communication in an SoC. The bugs and the faults which affect the communication between the cores are considered. We propose an online debug infrastructure for transaction-based debugging. Also we suggest a unified debug flow which takes the approaches of Parts I and II into account.

A brief presentation of the content of each remaining chapter concludes the introduction. The related work is discussed in each chapter according to the focus of the chapter. For convenience, acronyms and symbols used throughout the book are summarized on pages xi and xiii, respectively.

- Chapter 2 presents basic information about circuits and sensitized paths. This chapter deals with how a circuit is converted into *Conjunctive Normal Form* (CNF) as a suitable representation for a SAT-solver. Also, three-valued logic in SAT is described. Three-valued logic is used in formal verification and debugging to create strong counterexamples. Then the chapter describes SAT-based debugging for combinational and sequential circuits and shows how a circuit is enhanced using correction blocks in order to debug the circuit. The constraints required in CNF to achieve the debug automation are explained. Then the transaction elements and the *Transaction Debug Pattern Specification Language* (TDPSL) are introduced. TDPSL has three layers: Boolean layer, temporal layer, and verification layer. These layers are explained in this chapter.

1.1 Part I: Debug of Design Bugs

- Chapter 3 proposes an approach for automating the design debugging procedures by integrating SAT-based debugging with testbench-based verification. The diagnosis accuracy increases by iterating debugging and counterexample generation, i.e., the total number of fault candidates decreases. The approach uses diagnostic traces to obtain more effective counterexamples and to increase the diagnosis accuracy. The approach is utilized as a basis for the next chapter to develop a generalized debug flow.
- Chapter 4 presents a generalized approach to automate debugging which can be used in different scenarios from design debugging to post-silicon debugging. The approach is based on MBD. Diagnostic traces are proposed as an enhancement reducing debugging time and increasing diagnosis accuracy. This chapter also presents a concrete instantiation of the proposed generalized debugging as an automated approach for post-silicon debugging of design bugs by integrating post-silicon trace analysis, SAT-based debugging, and diagnostic trace generation. Our approach uses trace buffers as a hardware structure for debugging.
- Chapter 5 presents an approach to automatically debug synchronization bugs due to coding mistakes at RTL. In particular, an appropriate bug model is introduced and it is shown how synchronization bugs are differentiated from other types of bugs by our approach.

1.2   Part II: Debug of Delay Faults

- Chapter 6 proposes a methodology to model and evaluate the functional behavior
  of logic circuits under timing variation. In the approach, first a *Time Accurate
  Model* (TAM) of the circuit is constructed. The TAM represents the behavior
  of the circuit in the functional domain with respect to the circuit delay and
  a precision of an arbitrarily fine-grained but discrete time unit. The timing
  variations are modeled and are added to the TAM. Then the model is evaluated
  by analyzing logic circuits under timing variations.
- Chapter 7 utilizes the TAM and the timing variation model presented in Chap. 6
  to debug failing speedpaths which are frequency-limiting critical paths affecting
  the performance of a chip. Given an erroneous trace obtained from a testbench,
  a debug instance is created and is constrained to the erroneous trace in order to
  automatically diagnose failing speedpaths.
- Chapter 8 integrates STA and functional analysis in order to efficiently construct
  a compact timing model of a circuit. By using the STA, potentially-critical
  sections of the circuit are selected. Then timing variation models are built only
  for the potentially-critical sections of the circuit resulting in a decrease of the
  size of the debugging instance. Consequently, the debugging time decreases
  significantly.

1.3   Part III: Debug of Transactions

- Chapter 9 presents an approach for transaction-based online debugging of
  multiprocessor SoCs using a *Network-on-Chip* (NoC). Our approach utilizes
  monitors and filters as hardware structure. Monitors and filters observe and
  filter transactions at run time. They are connected to a *Debug Unit* (DU). Then,
  transaction-based programmable *Finite State Machines* (FSMs) in the DU are
  used as online assertions to assert the correct relation of transactions at run time.
  In this chapter the FSMs and filters are programmed according to race, deadlock,
  and livelock debug patterns. Also this chapter presents a debug flow using the
  proposed infrastructure which shows the application of the debug approaches of
  Parts I and II in the appropriate debug step.
- Chapter 10 summarizes the contributions presented in this book. Furthermore,
  possible directions of future research are mentioned.

Chapter 2
Preliminaries

2.1 Circuits and Sensitized Paths

Each combinational circuit is represented by a directed acyclic graph $C = (V, E)$, referred to as the *circuit graph*, where V is the set of circuit nodes and $E \subseteq V \times V$, the set of edges, corresponds to the gate input-output connections in the circuit [LRS89]. For gate-level benchmarks, we consider the nodes to be gates with symmetric functions. Each node in the circuit graph is associated with a symmetric function which represents the corresponding behavior of that gate in the circuit. A symmetric function does not depend on the order of inputs but only on the sum of variables assigned to 0 or to 1, respectively.

The basic gates used in this book for gate-level benchmarks are BUF, NOT, AND, OR, XOR, NAND, NOR and XNOR corresponding to the logic operators. Where relevant a FANOUT-gate may be inserted to differentiate between multiple fanout-branches of a gate. A FANOUT gate implements the identity function. For the RTL benchmarks, in addition to the logic operators, the word-level operators such as arithmetic operations (e.g. ADD, MUL, DIV), control and relational operations (e.g. ITE, LEQ, EQ) and array operations (e.g. READ, WRITE) are used [SKF+09]. In these cases the function associated to a node is not symmetric. Thus, the order of inputs to an operator is relevant to determine the function of the circuit. However, to simplify the presentation we do not reflect this in the following definitions. In the examples and explanations of this book, for the sake of simplicity, we use gates to explain the idea of our approaches.

The successors of a node $g \in V$ are given by a set of nodes $succ_g = \{h | (g, h) \in E\}$ and $|succ_g|$ is the number of successors of g. The predecessors of a node $g \in V$ are given by a set of nodes $pred_g = \{h | (h, g) \in E\}$ and $|pred_g|$ is the number of predecessors of g.

Gates are usually denoted by lower case latin letters in the circuit representation, e.g., gate a in Fig. 2.1. Letter I indicates the set of *Primary Inputs* (PI) and O is the set of *Primary Outputs* (PO). The PIs of a circuit are often denoted by $i_1, i_2, \ldots, i_n$

© Springer International Publishing Switzerland 2015
M. Dehbashi, G. Fey, *Debug Automation from Pre-Silicon to Post-Silicon*,
DOI 10.1007/978–3-319-09309-3_2

Fig. 2.1 An example circuit

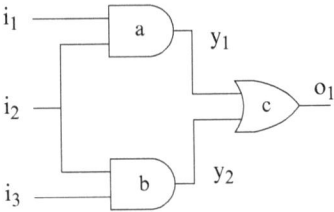

Fig. 2.2 Circuit graph

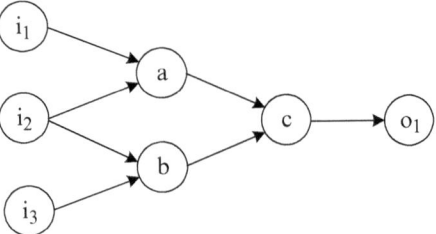

while the POs are often denoted by $o_1, o_2, \ldots, o_m$. The wires in the circuit are also denoted by lower case latin letters. Additional indices can be used for ordering wires and gates.

An example combinational circuit and its corresponding circuit graph are shown in Figs. 2.1 and 2.2, respectively. The circuit has three primary inputs which are shown by letters i_1, i_2 and i_3. Also there is one primary output o_1. In the circuit graph, each intermediate node represents a gate in the circuit. The source nodes, as the nodes which have only outgoing edges in the circuit graph, represent primary inputs. The primary outputs are represented by sink nodes, as the nodes which have only incoming edges in the circuit graph.

For symmetric gates, an input to a node is said to have a *controlling value (cv)* if it determines the value of the node output regardless of the values on the other inputs to the node. If the value on some input is the complement of the *cv*, the input is said to have a *non-controlling value (ncv)*. An input with value X does neither have a *cv* nor a *ncv*. Figure 2.4 shows the truth table of AND and OR gates. For an AND gate the value 0 is *cv* and the value 1 is *ncv*. For an OR gate, the value 1 is *cv* and the value 0 is *ncv*. For gates XOR and XNOR, no *cv* and *ncv* are defined because the output of gates XOR and XNOR always depends on the values of both inputs of the gate (in the case of two inputs). Therefore, if one of the inputs of gates XOR and XNOR has the value X (denoting an unknown value in three-valued logic), the output value will be X.

A path P from node g_1 to node g_r is a sequence of nodes $(g_1, g_2, \ldots, g_r)$. An edge as a wire in the circuit can have value 0 or 1 in the binary encoding. An edge whose value changes under the presence of some fault(s) is called a *sensitized edge*, and a path of sensitized edges is called a *sensitized path* [VH99]. In the ternary encoding, an edge can have value 0, 1 or X.

In contrast to a combinational circuit, a sequential circuit includes flipflops or state elements (Fig. 2.3). A flipflop is used to store state information in the circuit.

Fig. 2.3 Sequential circuit

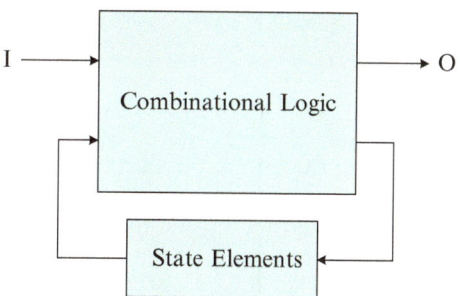

In a sequential circuit, output values of the circuit depend on the input values and the current state of the circuit (values of all flipflops). In general, the next state of the circuit is determined by the input values and the current state.

For sequential circuits, we denote a flipflop as $b = FF(a, clk, init)$, where a is the data input of the flipflop, clk indicates the clock frequency of the corresponding flipflop, and $init$ is the initial state of the flipflop. Wires a and clk and initial state $init$ are represented by Boolean variables with the same name in our models. The data output of the flipflop is denoted by b. A flipflop can change its state on either rising clock edge or falling clock edge. When a circuit has a single clock and the initial states of flipflops are 0, a flipflop is denoted as $b = FF(a)$.

2.2   Conjunctive Normal Form

Boolean Satisfiability (SAT) has emerged as a powerful reasoning engine for verification [Gup07] and also related areas, e.g., debugging [FSBD08] as well as diagnosis [SVAV05]. A SAT solver is applied to the logic formula of a circuit which has been converted into *Conjunctive Normal Form* (CNF). A CNF Φ on n Boolean variables $x_1, x_2, \ldots, x_n$ is a conjunction (AND) of m clauses $\omega_1, \omega_2, \ldots, \omega_m$. Each clause is the disjunction (OR) of one or more literals. A literal is a Boolean variable (x) or its complement $(\overline{x})$. The CNF as a Boolean formula is written as follows: $\Phi = f(x_1, x_2, \ldots, x_n)$.

The task of a SAT solver is to find an assignment $a_1, a_2, \ldots, a_n$ for $x_1, x_2, \ldots, x_n$ such that $f(a_1, a_2, \ldots, a_n) = 1$ or to prove that no such assignment exists. The Boolean formula f is called *satisfiable* (SAT) if such an assignment exists. Otherwise, f is called *unsatisfiable* (UNSAT).

For a circuit $C(V, E)$, where circuit nodes and edges are gates and their connections, respectively, CNF Φ_C has to be derived. To derive CNF Φ_C, a Boolean variable x_e is assigned to each connection $e \in E$. Then, for each gate $g \in V$, CNF Φ_g is derived from the characteristic function [Tse68, Lar92]. The characteristic function of a gate denotes whether a respective assignment to the gate inputs

Table 2.1 CNF formulas for simple gates [MS95]

Gate type g	Gate function	Φ_g
AND	$y = AND(w_1, \ldots, w_j)$	$\prod_{i=1}^{j} (w_i + \bar{y}) \cdot \sum_{i=1}^{j} (\bar{w}_i + y)$
NAND	$y = NAND(w_1, \ldots, w_j)$	$\prod_{i=1}^{j} (w_i + y) \cdot \sum_{i=1}^{j} (\bar{w}_i + \bar{y})$
OR	$y = OR(w_1, \ldots, w_j)$	$\prod_{i=1}^{j} (\bar{w}_i + y) \cdot \sum_{i=1}^{j} (w_i + \bar{y})$
NOR	$y = NOR(w_1, \ldots, w_j)$	$\prod_{i=1}^{j} (\bar{w}_i + \bar{y}) \cdot \sum_{i=1}^{j} (w_i + y)$
NOT	$y = NOT(w_1)$	$(y + w_1) \cdot (\bar{y} + \bar{w}_1)$
BUFFER	$y = BUFFER(w_1)$	$(\bar{y} + w_1) \cdot (y + \bar{w}_1)$

and output is consistent or inconsistent with the gate's function. The CNF Φ_C is constructed by the conjunction of the CNF of each gate:

$$\Phi_C = \prod_{g_i \in V} \Phi_{g_i} \tag{2.1}$$

The number of clauses in Φ_C is linearly related to the number of circuit nodes if only gates up to a fixed number of inputs or no XOR/XNOR-gates are considered. Table 2.1 shows the CNF formulas for simple gates. For each simple gate with j inputs, $j + 1$ clauses are created. For example, for a two-input AND gate $y = w_1 \cdot w_2$, the CNF formula is given by the following expression:

$$\Phi_{AND} = (w_1 + \bar{y}) \cdot (w_2 + \bar{y}) \cdot (\overline{w_1} + \overline{w_2} + y) \tag{2.2}$$

The CNF formula for the example circuit of Fig. 2.1 is given by the following expression in which the CNF formula is created from Table 2.1 for each gate:

$$\begin{aligned} \Phi_C = {} & (i_1 + \overline{y_1}) \cdot (i_2 + \overline{y_1}) \cdot (\overline{i_1} + \overline{i_2} + y_1) \cdot \tag{2.3} \\ & (i_2 + \overline{y_2}) \cdot (i_3 + \overline{y_2}) \cdot (\overline{i_2} + \overline{i_3} + y_2) \cdot \\ & (\overline{y_1} + o_1) \cdot (\overline{y_2} + o_1) \cdot (y_1 + y_2 + \overline{o_1}) \end{aligned}$$

In *Bounded Model Checking* (BMC) the behavior of a sequential circuit is modeled in CNF. In BMC, a sequential circuit is unrolled for some time steps (clock cycles). In this case, the flipflops are removed and the input wire of a flipflop from clock cycle i is connected to the appropriate gates in clock cycle $i + 1$. In each unrolled copy of the circuit, only the combinational logic remains. Then the whole CNF is created by the conjunction of the CNFs of all unrolled copies.

2.3 Three-Valued Logic in Boolean Satisfiability

In three-valued logic, each variable can have the value 0, 1, or don't care X. This logic has been used in the field of formal hardware verification for creating strong counterexamples [RS04, GK05] and faster verification engines [Vel05, SVD08]. Also, this logic has been used for generating high quality counterexamples [SFB+09].

An encoding in *Conjunctive Normal Form* (CNF) is needed for the three-valued logic defined over {0, 1, X} to apply a standard SAT solver. Accordingly, the modeling of gates and components in the CNF formula has to be adjusted. Here, three-valued logic is encoded by using two variables for each signal similar to [Vel05]. The three-valued constants 0, 1, and X are defined by pairs $(0, 0)$, $(0, 1)$ and $(1, -)$, respectively. The first variable of a pair determines whether the signal has a binary value or value X. If the first variable is 0, the signal has binary value. In this case, the second variable shows a binary value as 0 or 1. If the first variable is 1, the signal has value X. In this case, the second variable is not used. Figure 2.4 shows the truth table of AND and OR gates regarding three-valued logic. When one input has a ncv, value X can be propagated to the gate output.

2.4 Model-Based Diagnosis

MBD is an approach to diagnose errors from first principles [Rei87, FSW99, MS07]. In MBD, a system model is provided in terms of components and their interconnections [Rei87, dKK03]. The component models describe how each component behaves. The behavior of the system is modeled as a logic formula. Then a domain-independent reasoning engine calculates the diagnosis from the model and system observations. Independently, SAT-based reasoning engines have been shown as a robust and efficient approach to diagnose faults and localize the bugs, called SAT-based debugging [SVAV05]. The underlying principle of MBD and SAT-based debugging is very similar. In the following section, SAT-based debugging is explained.

AND	0	1	X		OR	0	1	X
0	0	0	0		0	0	1	X
1	0	1	X		1	1	1	1
X	0	X	X		X	X	1	X

Fig. 2.4 Truth table of AND and OR gates with three-valued logic

2.4.1 SAT-Based Debugging

Debugging is a procedure in a design process that is started when the implementation of the design has failed verification. The output of the verification engine is typically returned as a set of counterexamples which proves the existence of a bug in the implementation. The set of counterexamples is denoted by CEs. Each counterexample is denoted by $CE_i \in CEs$. A counterexample $CE_i \in CEs$ includes input stimuli causing erroneous behavior, and the expected correct output response, i.e., $CE_i = (I_i, O_i)$. An input stimulus (input test vector) is a single assignment of the PIs $I = (i_1, i_2, \ldots, i_n)$ and is denoted by $(v_1, v_2, \ldots, v_n)$ where $v_i \in \{0, 1, X\}$. When sequential behavior is considered a superscript or subscript may be used to denote time points.

A circuit is generally divided into components. The components specify the granularity of the debugging result. Typical components are gates or modules, but also the hierarchical and structural information can be taken into account [ASV+05,FD05,FSBD08,SVAV05]. We consider gates as components in examples of this book for the sake of simplicity.

An approach for SAT-based debugging was presented in [SVAV05] that searches for all possible fault candidates in the implementation. Each fault candidate is a set of components of the implementation which may be modified to fix the erroneous behavior of the counterexamples. Firstly, the circuit is enhanced with correction blocks by adding a multiplexer at the output of each component. The original output function f_c of component c is replaced by f_c' as shown in Fig. 2.5a. The select line sel_c of the added multiplexer controls f_c' such that if sel_c is activated $f_c' = r_c$ where r_c is an unconstrained variable and a value for correcting the erroneous behavior may be injected, otherwise $f_c' = f_c$. The select line is also called *abnormal predicate*.

Given an implementation of a circuit and a set of counterexamples, one copy of the circuit is created for each counterexample. Then, the inputs and outputs are constrained to the input stimuli and to the correct output response of the corresponding counterexample as shown in Fig. 2.5b. When abnormal predicates are inactive, the created problem instance yields a contradiction as the circuit produces erroneous output. In Fig. 2.5b,c for the sake of simplicity, only the select lines of the correction block are shown. In the figures, the block "+" is a binary adder with m one-bit inputs [SVAV05].

When debugging is started, the SAT solver searches for satisfying assignments by activating some of the abnormal predicates. The number k of active abnormal predicates is controlled by a fault cardinality constraint as shown in Fig. 2.5b. The debugging procedure increases the number k from 1 until a satisfiable instance is found. Here, the number k shows the minimal number of changes leading to a satisfiable instance.

Debugging for sequential circuits is done by unrolling the circuit for some time steps equal to the length of the counterexample [FAVS+04] as shown in Fig. 2.5c. The length of the counterexample is the number of clock cycles in which the input

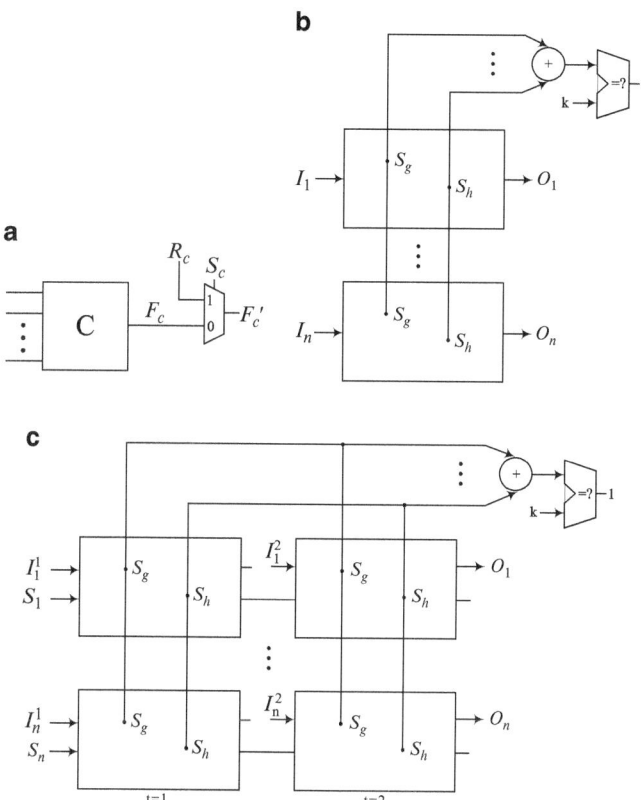

Fig. 2.5 SAT-based debugging. (**a**) Correction block. (**b**) Combinational debugging. (**c**) Sequential debugging

stimuli create the erroneous output value. In the example of Fig. 2.5c, the length of counterexamples is two time steps. The correction block is added as in the combinational case and usually the same abnormal predicate is used for the same gate in all time steps and for all counterexamples. The set of the states of the circuit (flipflops) is denoted by S. An index is used to show the time step of the states.

The case in which one component in the circuit is faulty is called *single fault*. In the case of *multiple faults*, more components than one are faulty in the circuit. Conceptually, for each counterexample CE_i, $i = 1, 2, \ldots, n$, where n is the number of counterexamples, there is a set of fault candidates $\mathscr{F}_{CE_i}$. In the case of single faults, these sets are intersected to return the final set of fault candidates $\mathscr{F}$ regarding all counterexamples:

$$\mathscr{F} = \mathscr{F}_{CE_1} \cap \mathscr{F}_{CE_2} \cap \ldots \cap \mathscr{F}_{CE_n} = \bigcap_{i=1}^{n} \mathscr{F}_{CE_i} \qquad (2.4)$$

The set $\mathscr{F}$ contains a set of components of the circuit. In the examples of this book, the set $\mathscr{F}$ is a set of gates. Each fault candidate is denoted by $FC_i \in \mathscr{F}$. Parameter n has a direct impact on the memory and the run time of debugging engines. A debug algorithm which can minimize the number of fault candidates with a minimum number of counterexamples in a reasonable time improves the performance of the debugging engines.

Diagnosis accuracy is a function of the fault candidates $\mathscr{F}$. In general, the quantity and quality of fault candidates determine the diagnosis accuracy. For design bugs, we use the number of fault candidates as an estimate for diagnosis accuracy. In this case, a smaller number of fault candidates indicates a higher diagnosis accuracy. For delay faults, not only the number of fault candidates, but also the number of paths constituted by fault candidates and also the length of the shortest path from a real fault location to the fault candidates are used to indicate the diagnosis accuracy. A smaller number of paths with a smaller number of gates on them with a shorter distance from a real fault location indicates a higher diagnosis accuracy.

For multiple faults, each fault candidate $FC_i \in \mathscr{F}$, $i = 1, 2, \ldots, |\mathscr{F}|$, is a tuple of k components, $FC_i = \{FC_{i,1}, FC_{i,2}, \ldots, FC_{i,k}\}$. $FC_{i,j}$ is the jth component of the fault candidate FC_i. The set $\mathscr{F}$ is constituted in a way that each component of a fault candidate FC_i is part of at least one sensitized path of one counterexample. Each counterexample implies one or more sensitized paths. The gates along the sensitized paths are the potential places on which the fault may be located. When there are multiple counterexamples and in each counterexample all faults are sensitized, the sensitized paths of counterexamples have common gates which define their intersection. Because the sensitized paths are derived from the faulty components. In this case, the gates on the intersected sensitized paths are potential components of fault candidates which propagate the error through different paths. All components of a fault candidate can propagate the error through all sensitized paths determined by the counterexample. If one fault is sensitized and another fault is not sensitized in a counterexample, the sensitized paths of counterexamples for multiple faults may not have an intersection.

Fault model-free SAT-based debugging does not make any assumption on the fault types present in the circuit for the both cases of single fault and multiple faults. The advantage is that faults with a nondeterministic (unmodeled) behavior can also be captured [Ait97, LV05]. As shown in Fig. 2.5a, there is no restriction on the correction variable r_c. When select line sel_c is activated, the SAT solver can assign any value to the correction variable r_c to justify the erroneous behavior of the outputs corresponding to the counterexample. Since there is no restriction on the correction variables, the fault model-free debugging is performed [Ait97]. The correction blocks used for model-free debugging can change the function of each component in the circuit. In this case, local correction regarding one component is performed and there is no assumption about the function of a top module which contains the component having an activated correction block.

In the case of multiple faults, fault masking may occur in which one fault (dominant fault) masks the behavior of another fault (masked fault). In this case, the behavior of the masked fault is not propagated to an output and only dominant fault

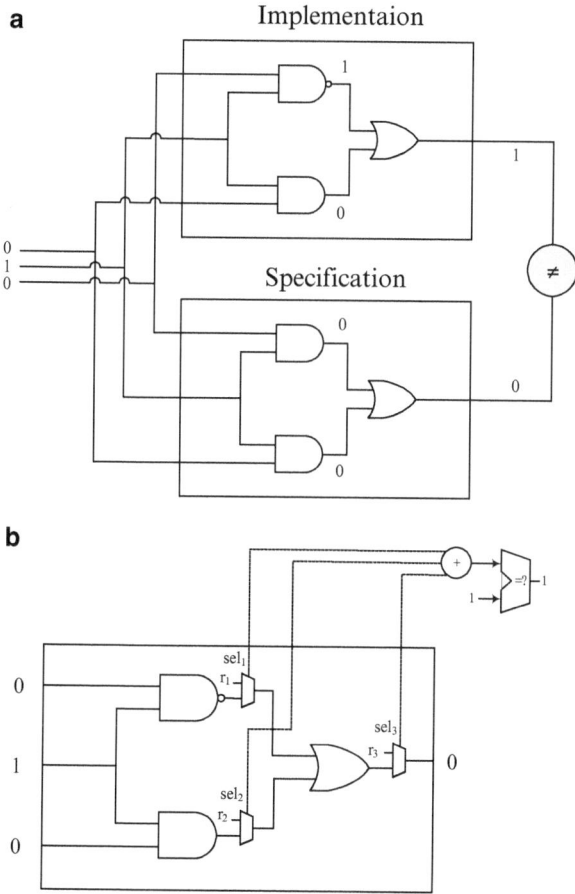

Fig. 2.6 Verification and debugging example. (**a**) Verification. (**b**) Debugging instance

is found as fault candidate. The masked fault is missed. This is a known problem but not addressed in this book. The quality of counterexamples may influence the fault masking. If a counterexample can activate different sensitized paths for different faults such that the behavior of each fault can be propagated to outputs through at least one sensitized path, the fault masking does not occur.

Example 2.1. Figure 2.6 shows the verification and the debugging processes for an implemented circuit. In the verification process, the implemented circuit (implementation) is compared to the specification. Here, the specification is given as a formal specification. If there is any inconsistency between the specification and the implementation, this inconsistency is considered as a counterexample. For instance in Fig. 2.6a, when the input vector is $(i_1, i_2, i_3) = (0, 1, 0)$, the outputs of the implementation and the specification are different. In this case, the input test vector and the correct output value obtained from the specification constitute a counterexample, i.e., $CE = (I, O) = (0, 1, 0, 0)$.

The counterexample *CE* is given for debugging. Debugging finds the fault candidates according to the given counterexample. To debug the implementation, first the implementation is enhanced by correction blocks to create the debugging instance (Fig. 2.6b). One correction block is inserted at the output of each gate. The select lines of the correction blocks are controlled by the fault cardinality constraint.

The debug instance is translated into CNF. A SAT solver is utilized to find and to enumerate all possible solutions (fault candidates). The SAT solver finds the following solutions: $(sel_1 = 1, r_1 = 0)$, $(sel_3 = 1, r_3 = 0)$. In this case, the NAND and OR gates are considered as fault candidates, as their corresponding multiplexers are the solutions found by the SAT solver. The solution $(sel_1 = 1, r_1 = 0)$ means if the select line sel_1 is activated and the correction value $r_1 = 0$ is inserted in the circuit, the correct output value is created. Accordingly, the solution $(sel_3 = 1, r_3 = 0)$ means that activating the select line sel_3 and injecting the correction value $r_3 = 0$ through the input of the multiplexer leads to the correct output value.

Figure 2.7 shows the algorithm of SAT-based debugging as pseudocode. The inputs of the algorithm are an implemented design and a set of counterexamples. In line 2, the design is copied as many times as the number of counterexamples. Also, the inputs and the output of each copy is constrained according to the inputs and the output values of one counterexample. Then the correction blocks are inserted in the model (line 3). The set of fault candidates $\mathscr{F}$ is initially empty (line 5) and variable k is initialized to 1 (line 6). The select lines of multiplexers are constrained according to k in line 9. If the SAT solver can find a solution (line 11), the solutions

```
1   algorithm  SAT_Based_Debugging (In : Design,CEs,  Out : F)
2   Copy_Constrain (Design,CEs)
3   SEL = Insert_Correction_Block()
4   // sel_i ∈ SEL, i = 1, 2, . . . , m
5   F = ∅
6   k = 1
7   do
8   {
9       Add_Constraint ( (∑_{i=1}^{m} sel_i) = k )
10
11      if  Solve() == SAT  then
12      {
13          F = Extract_All_Solutions()
14          break
15      }
16      else
17      {
18          Remove_Constraint ( (∑_{i=1}^{m} sel_i) = k )
19          k = k + 1
20      }
21  } while  k ≤ |SEL|
22  end function
```

Fig. 2.7 SAT-based debugging

are returned as fault candidates (line 13) and the algorithm terminates. Otherwise, the previous constraint is removed (line 18) and variable k increases (line 19). The algorithm repeats until variable k reaches the total number of select lines (line 21). In the worst case, the loop in the algorithm is executed $|SEL|$ times, where $|SEL|$ is the number of components in the circuit.

2.5   Transaction-Based Debug

In this section, the basic information about transaction-based debugging is explained. Transaction based communication-centric debugging was introduced to debug complex and large SoCs [GVVSB07]. Transaction-based debugging has more coarse-grain diagnosis accuracy in comparison to debugging at the gate level because in transaction-based debugging only transactions and events are considered. However, the approach is especially suitable for large SoCs. The output of transaction-based debugging can be utilized when debugging at RTL and gate level in order to find the root cause of the error. In an SoC, there are some IP cores which interact through a network. The transactions are observed using monitors [GVVSB07] and a debug unit controls the execution of the SoC. Also, the transactions are monitored online and are stored in on-chip memories such as a trace buffer. After running the SoC, the transaction-based assertions are used to find certain patterns in the extracted transactions [GF09]. *Transaction Debug Pattern Specification Language* (TDPSL) is used to define the transaction-based assertions. As the name suggests TDPSL is derived from *Property Specification Language* (PSL) [IEE05]. In the following, the concept of transaction and the TDPSL are explained.

2.5.1   Transaction

This section explains the transaction elements from [Ope09] and [GF09]. In TLM, abstract operations which describe the communication in a system are called transactions. Each transaction consists of a request from a master and a response from a slave. More precisely, a transaction in TLM has four basic elements:

- *Start of Request (SoRq)* corresponds to putting the request in the channel (communication medium) by the master.
- *End of Request (EoRq)* is getting the request from the channel by the slave.
- *Start of Response (SoRp)* corresponds to putting the response in the channel by the slave.
- *End of Response (EoRp)* is getting the response from the channel by the master.

Fig. 2.8 Elements of a
transaction

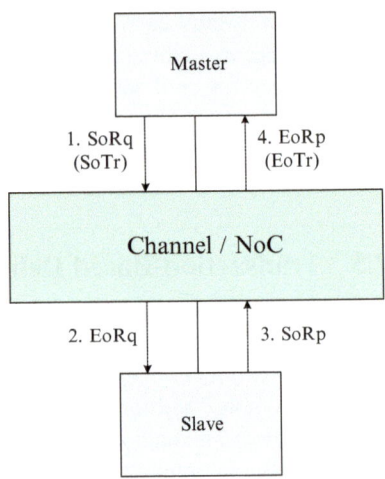

Figure 2.8 shows the basic elements of a transaction which are transferred via a channel or an NoC. Also there are two additional elements which are called: *Request Error* (ErrRq) and *Response Error* (ErrRp). These elements handle error conditions and correspond to any kind of error that causes a request or a response to fail.

2.5.2 Transaction Debug Pattern Specification Language

TPDSL has three layers as shown in Fig. 2.9: the Boolean layer, the temporal layer, and the verification layer [GF09]. The Boolean layer contains *trans_exp* which represents the basic elements of transactions. The *trans_exp* format is as follows:

   trans_type (*master, slave, type, address, tag*)

Field *trans_type* can be any transaction element mentioned in Sect. 2.5.1 as well as the *Start of Transaction* (SoTr) or the *End of Transaction* (EoTr) which are the same as SoRq and EoRp, respectively. The elements SoTr and EoTr model transactions more coarsely than the elements SoRq, EoRq, SoRp and EoRp. Fields *master* and *slave* specify the master and slave IDs which are an abstraction of the higher address bits. Field *type* can be read Rd or write Wr. Field *address* indicates the slave address symbolically as SAME, SEQ, and OTHER. Field *tag* indicates the transaction number and is only used for buses that allow non-blocking requests and out-of-order responses [GF09]. In this book, we show a transaction without considering the field *tag*.

   The motivation to use symbols for the address field is to abstract and to compress the address bits. In this case, only the compact address information is stored or sent via network for debugging. The symbols can be defined with respect to the application and the granularity of debugging. The following symbols are used: SAME specifies that in the current transaction, the slave address is the same as

```
/********** Boolean Layer **********/
trans_exp ::=
    trans_type "(" master "," slave "," type "," address "," tag ")";
trans_type ::=
    "SoRq" | "EoRq" | "SoRp" | "EoRp" | "ErrRq" | "ErrRp" |
    "SoTr" | "EoTr" | "ErrTr" ;
master ::= var | id | "-" ;
slave ::= var | id | "-" ;
tag ::= var | id | "-" ;
type ::= "Rd" | "Wr" | "-" ;
address ::= "SAME" | "SEQ" | "OTHER" | "-" ;
var ::= IDENTIFIER ;
id ::= NUMBER ;

/********** Temporal Layer **********/
fl_exp ::= ("always" | "never" | "eventually") temporal_exp ;
temporal_exp ::= seq_exp | temporal_exp t_op temporal_exp ;
t_op ::= "->" | "&&" | "||" ;
seq_exp ::= sere_exp | "{" sere_exp "}" | seq_exp repeat_exp;
repeat_exp ::= repeat_op [ count_exp ] ;
repeat_op ::= "*" | "+" | "=" ;
count_exp ::= NUMBER [ ":" (NUMBER | "inf") ] ;
sere_exp::= trans_exp | sere_exp sere_op sere_exp ;
sere_op ::=";" | ":" | "|" | "&" ;

/********** Verification Layer **********/
property_exp ::= "assert" fl_exp [ filter_exp ] ;
filter_exp ::= "filter" "(" masters "," slaves "," types ")" ;
masters ::= "*" | "-" | id_list ;
slaves ::= "*" | "-" | id_list ;
types ::= "*" | type ;
id_list ::= id | id_list "&" id ;
```

Fig. 2.9 The syntax of TDPSL [GF09]

the previous transaction address for this slave. SEQ specifies that in the current transaction, the slave address has one word difference with the previous transaction address for this slave. OTHER specifies that in the current transaction, the slave address is neither SAME nor SEQ.

Example 2.2. Transaction *EoTr* ($m1$, $s2$, Rd, $-$) represents the end of a read transaction from master $m1$ to slave $s2$ with any address. The symbol "$-$" indicates that we leave the address field as don't care. As explained before, the tag field is not shown.

The properties in terms of transaction sequences are defined at the temporal layer. Different operators are available at this layer such as *concatenation operator* (;), *fusion operator* (:), *or operator* (|), *and operator* (&) and *repetition operators*

$(*, +, =)$ [GF09]. The *always* operator means that the temporal expression has to hold at any time. The *never* operator means that the temporal expression has to never hold. The *eventually* operator means that the temporal expression has to hold at the current time or some future time [GF09].

Last layer in Fig. 2.9 is verification layer. In the verification layer, the *assert* statement is defined. Also, a *filter* can be defined which specifies a filter over the execution path for the evaluation of the assertion statement. The filter expression *filter_exp* in the verification layer of Fig. 2.9 can be used to filter some of the unwanted transactions. Filters can be defined over masters, slaves and transaction types. Filters can be defined by explicitly naming the masters, slaves or transaction types. The symbol "−" means no filter and "∗" means that only consider related masters, slaves or transaction types [GF09]. In the following example, the related masters are $m2$ and $m1$. The related slave is $s1$.

Example 2.3. A simple assertion in TDPSL is as follows:
   assert never{
      EoTr $(m2, s1, Wr, -)$; *SoTr* $(m1, s1, Rd, -)$
   }*filter*$(*, *, *)$

This assertion specifies that start of a read transaction from master $m1$ to slave $s1$ must never be directly after the end of a write transaction from master $m2$ to slave $s1$. Sign ∗ in the filter means only the related transaction types, masters, and slaves have to be considered.

For SoCs, debugging can be performed at the granularity of transaction-level and message-level [GVVSB07]. For transaction-level debugging, we need to observe each transaction including SoTr and EoTr. As Fig. 2.8 shows, SoTr and EoTr can be observed at master interconnect. Monitors on master interconnects can be utilized to observe transactions (Fig. 2.10b). Message-level debugging has a finer granularity. A message is a request or a response. Typical requests are read and write commands. Read data and write acknowledgment are typical responses [GVVSB07]. To perform message-level debugging, we need to observe each message including

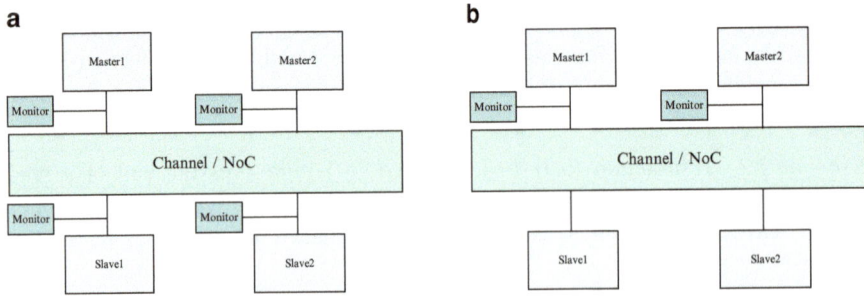

Fig. 2.10 Monitoring. (**a**) Message-level monitoring. (**b**) Transaction-level monitoring

SoRq and EoRq as well as SoRp and EoRp. As Fig. 2.8 shows, both master and slave interconnects have to be observed for message-level debugging. Monitors on both master and slave interconnects can be utilized to observe the messages (Fig. 2.10a).

2.6   Overview of Benchmarks

In this book, we propose automated debugging approaches for a hardware system at different abstraction levels, i.e., transaction-level, RTL and gate-level. To show the efficiency of our approaches, we use suitable benchmarks at each abstraction level. Figure 2.11 shows the structure of our benchmarks for debugging at different levels. As Fig. 2.11 shows, at transaction-level we use an NoC-based SoC. In this case, we utilize the Nirgam environment which is a cycle accurate NoC simulator implemented in SystemC language [Jai07]. In Fig. 2.11, the system consists of *Routers* (R) and *Intellectual Property* cores (IP). Each IP has two main elements: *Processing Element* (PE) and *Communication Element* (CE).

The elements of an IP are considered for debugging at RTL. In this case, we consider the OpenRISC processor [LB14], ITC-99 [Dav99] and LGSynth-93 [McE93] benchmark suites. Using ITC-99 and LGSynth-93 benchmarks also help us to compare our approaches with previous debugging approaches which will be mentioned in later chapters.

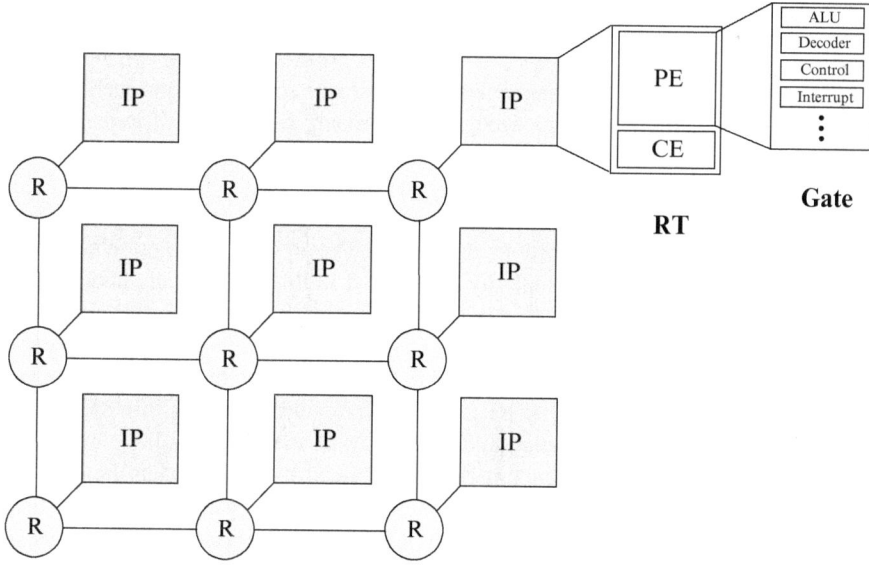

Fig. 2.11 Structure of debugging benchmarks at different levels: Transaction, RT, Gate

Table 2.2 Benchmarks

Debug of transactions	Debug of design bugs	Debug of electrical faults
NoC-Based SoC (Nirgam)	OpenRISC ITC-99 LGSynth-93	ISCAS-85 ISCAS-89

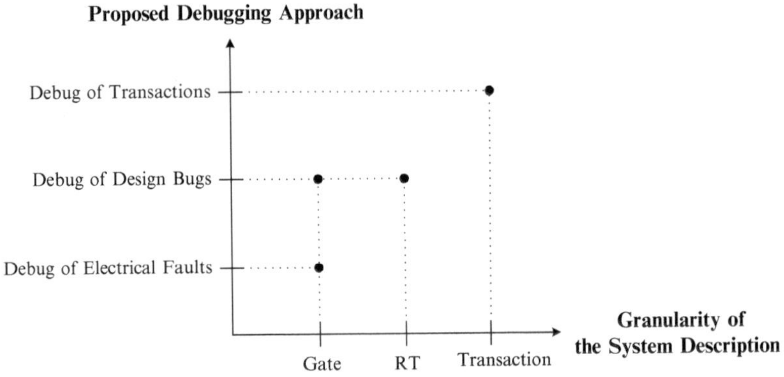

Fig. 2.12 The proposed debugging approaches regarding the granularity of the system description

A PE can have different modules such as ALU, decoder, control unit, interrupt controller, etc., which are represented by the ISCAS benchmarks [BPH85, Brg85, BBK89]. The ISCAS benchmarks are known benchmarks in the research area for test of delay faults. Also, we use the ISCAS benchmarks to evaluate our approaches to debug delay faults.

Table 2.2 summarizes the benchmarks used for our debugging approaches. The first row indicates the proposed debugging approach, i.e., debug of transactions, debug of design bugs and debug of electrical faults (delay faults). The second row indicates the used benchmark suite.

Figure 2.12 shows the proposed debugging approaches regarding the granularity of the system descriptions used in this book. The system description at gate-level is the most fine-grain granularity considered in this book. Many details of a hardware system are available at this level. We even include some technological parameters like timing behavior of gates in respective models. The most coarse-grain granularity of the system description in this book is at transaction-level.

For debug of electrical faults, we use the gate-level description of a system as certain information such as delay and frequency of the circuit is only available at gate-level. The gate-level benchmarks are obtained by synthesizing the circuits using Synopsys Design Compiler with Nangate 45nm Open Cell Library [Nan11]. For debug of design bugs, we use both RTL and gate-level descriptions of a system. Because a design bug may escape the verification at RTL and may slip into gate-level. The transaction-level description of a system is used for validating the methodology for transaction-based debugging.

Part I
Debug of Design Bugs

Chapter 3
Automated Debugging for Logic Bugs

Automated debugging approaches are necessary to speed up the design process as size and complexity of VLSI designs increase. Among these approaches, debugging based on SAT [SVAV05] has been shown as a robust and efficient approach. The purpose of SAT-based debugging is to identify the potential sources of an observed error by using the available counterexamples utilizing the practical efficiency of SAT-based reasoning engines for NP-complete problems. Each potential source of the error is returned as a fault candidate which is a set of components of the circuit. A fault candidate can fix all erroneous behavior of counterexamples using non-deterministic replacements.

Different approaches have been proposed for improving the performance and accuracy of SAT-based debugging. They can be categorized into two groups. The approaches of the first group enhance the debugging performance for a given set of counterexamples. The approaches of the second group generate more counterexamples to improve the accuracy of the debugging process.

The approaches based on a given set of counterexamples aim at reducing run time and memory requirements. The work in [ASV+05] exploits the hierarchical structure of modern designs to improve the performance of debugging. *Quantified Boolean Formulae* (QBF) are utilized to reduce the size of the instance in [MVS+07]. In [CSMSV10], *Maximum SATisfiability* (MaxSAT) [BHvMW09] improves the performance and applicability of debugging. MaxSAT allows for a simple formulation of the debugging problem and therefore reduces the problem size and run time. Abstraction and refinement techniques are used in [SV07] for handling large designs with a better performance and reduced memory consumption. Fault candidates are ranked in [JLJ09] for a given set of counterexamples. The main drawback of these approaches is that the diagnosis accuracy is limited by the given set of counterexamples.

The approaches of the second group combine counterexample generation (verification) and debugging in a single flow. With each new counterexample, the *diagnosis accuracy* increases by excluding fault candidates that cannot fix erroneous

© Springer International Publishing Switzerland 2015
M. Dehbashi, G. Fey, *Debug Automation from Pre-Silicon to Post-Silicon*,
DOI 10.1007/978–3-319-09309-3_3

behavior of the added counterexample. Thus, a high quality counterexample reduces the number of fault candidates effectively. The work in [CMB07b] uses randomly generated counterexamples for debugging and applies automatic correction based on re-synthesis. Automatic correction increases the computational cost and is not guaranteed to fix an error in the desired way. The diagnosis accuracy measured, e.g., by the number of potential bug locations or by the distance between potential bug locations and real bugs, may decrease using random counterexamples. Also, the number of iterations between verification and debugging may increase. In [SFB+09], a heuristic approach based on three-valued logic is used to find high quality counterexamples. For an injected *X value* at a fault candidate and by observing X values on outputs, the approach assumes that modifying the fault candidate can create any value at the output. But this is an over-approximation due to the conservative properties of X values. The work in [SFD10] proposes an exact approach based on QBF which creates high quality counterexamples to find fault candidates fixing any erroneous behavior. However, the explicit enumeration of fault candidates may decrease the debugging speed for large designs. The approaches in [SFB+09] and [SFD10] need a formal specification for creating high quality counterexamples. However, a formal specification is often not given for complex designs. Here, a testbench is used to create the expected output response of an input stimulus.

In this chapter, we present a flow to improve the accuracy of SAT-based debugging when a testbench is used for verification [DSF13]. No formal specification is required. At first diagnostic traces are derived from the faulty implementation. A *diagnostic trace* is an input stimulus which tries to activate a fault candidate (or a set of fault candidates) and to propagate its behavior to the outputs. The diagnostic traces help testbench-based verification in creating high quality counterexamples. Then these counterexamples are used for iterating SAT-based debugging and increasing the diagnosis accuracy. The heuristic techniques in this chapter do not need a fault model for generating the diagnostic traces. Moreover, diagnostic traces are created by way of the faulty implementation and the initial set of fault candidates only. Whereas the focus of this chapter is on designs using a testbench, the proposed techniques for generation of diagnostic traces can also be applied for debugging designs using a formal specification. We compare three heuristics to find diagnostic traces with respect to run time, memory, and accuracy in experimental results. Also, the proposed heuristics are compared to random trace generation and to the formal approach of [SFD10] that is exact but requires a formal specification.

The remainder of this chapter is organized as follows. Our approach is presented in Sect. 3.1. We propose different heuristic techniques to generate diagnostic traces for design bugs in Sects. 3.3, 3.4, and 3.5, respectively. Section 3.6 presents experimental results on benchmark circuits. Section 3.7 summarizes this chapter.

3.1   Integration of Formal Debugging with Testbench-Based Verification

In this section we explain how SAT-based debugging and counterexample generation are integrated in one flow. In the approach, the diagnostic traces help creating high quality counterexamples for automated design debugging to increase the diagnosis accuracy.

The overall approach which consists of three main steps is shown in Fig. 3.1. Debugging, diagnostic trace generation, and running the testbench to validate diagnostic traces are three main steps of the approach. Having the design and an initial counterexample, debugging is started for finding all fault candidates which can correct the erroneous behavior of the circuit exhibited by the initial counterexample.

The second step of the approach, called diagnostic trace generation, is the main focus of this work. The inputs of this step are the faulty design and the set of fault candidates. The goal of this step is to generate diagnostic traces by heuristic methods. As a high quality counterexample aims at reducing the number

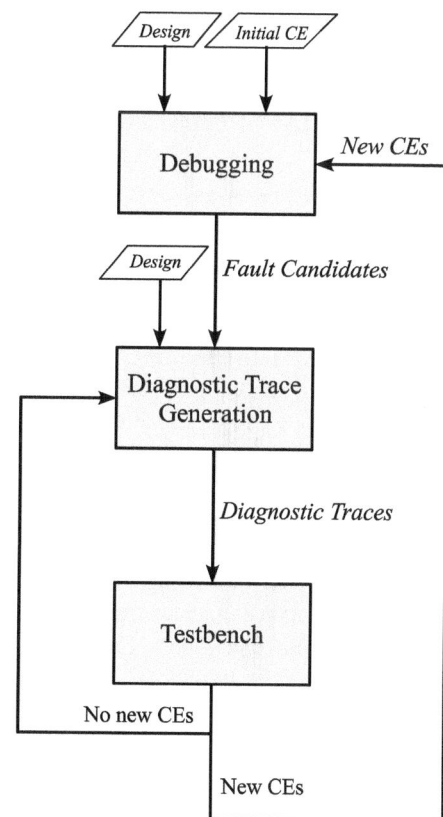

Fig. 3.1 Integration of debugging and testbench-based verification

of fault candidates effectively, each diagnostic trace activates a small number of fault candidates and observes their behavior on the outputs. Therefore, the counterexamples derived from diagnostic traces are likely to reduce the number of fault candidates.

Afterwards, the diagnostic traces are validated in a testbench environment, because it is not guaranteed that the diagnostic traces really create erroneous output responses in the design. Here, a testbench is used as a black box specification for creating the expected correct output of a diagnostic trace. A diagnostic trace creating erroneous output behavior is a counterexample and the step validates the diagnostic traces. If there is no counterexample, the algorithm returns to the second step for generating more diagnostic traces. Otherwise, the algorithm continues debugging with the new counterexamples. The number of iterations between debugging and verification is controlled by the diagnosis accuracy. The diagnosis accuracy is a function of the fault candidates, e.g., a small number indicates good accuracy. If the diagnosis accuracy is sufficient, the debugging algorithm terminates.

In the following, the intuition behind the diagnostic traces to create high quality counterexamples is described in Sect. 3.2. Three heuristics for diagnostic trace generation are introduced in Sects. 3.3, 3.4, and 3.5. In each section, first the discussion starts with a single fault assumption then the extension to multiple faults is explained, respectively.

3.2 Counterexample Versus Fault Candidate

In this section we describe why more counterexamples are effectively reducing the number of fault candidates and how to derive high quality counterexamples. Figure 3.2a represents a faulty circuit with a single fault. The real fault location is shown by a circle. The counterexample CE_1 is propagated through the dashed path to the circuit output o_1. The fault candidates indicated by $\times$ can correct the erroneous behavior of o_1. Actually, the fault candidates show the sensitized path related to the counterexample. The set of fault candidates related to CE_1 is denoted by $\mathscr{F}_{CE_1}$. Figure 3.2b shows the effect of the second counterexample CE_2 separately. The counterexample CE_2 is propagated through another path to o_2 and creates $\mathscr{F}_{CE_2}$. The effect of using both CE_1 and CE_2 in the debugging procedure is described in Fig. 3.2c. According to Formula (2.4), the number of fault candidates is reduced to the set of $\mathscr{F} = \mathscr{F}_{CE_1} \cap \mathscr{F}_{CE_2}$ where each fault candidate $FC_i \in \mathscr{F}$, $i = 1, 2, \ldots, |\mathscr{F}|$, is a single fault candidate. Now it is interesting to figure out how a high quality counterexample can have the strongest effect in reducing the number of fault candidates. Figure 3.2d shows the counterexample CE_3 which further reduces the size of $\mathscr{F}$. As shown in Fig. 3.2d, the sensitized path of CE_3 has the minimum intersection with the sensitized paths of other counterexamples.

In the case of multiple faults, each fault candidate FC_i can have k components. For this case, the sensitized paths of a new counterexample intersect with a fault candidate FC_i when all of the sensitized paths leading to erroneous behavior propagate

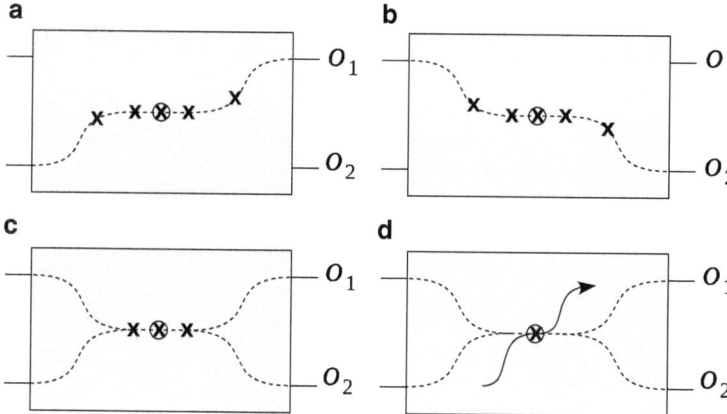

Fig. 3.2 Counterexample versus fault candidate. (**a**) Fault candidates of CE_1. (**b**) Fault candidates of CE_2. (**c**) Fault candidates of CE_1 and CE_2. (**d**) Fault candidates of CE_1, CE_2, and CE_3

through the components of FC_i. Thus, if a new counterexample includes sensitized paths that have minimum intersection with fault candidates, this counterexample can effectively reduce the size of $\mathscr{F}$.

For multiple faults, if all faults are activated in each counterexample, the sensitized paths of the counterexamples have an intersection because the sensitized paths are derived from the faulty components. If one fault is not sensitized in a counterexample for multiple faults, the sensitized paths of the counterexamples may not have an intersection. A more comprehensive discussion on multiple faults can be found, e.g., in [SFBD08].

3.3 Local Branch Activation

In this heuristic, the don't care value X is considered as a token for the behavior of one fault candidate and the algorithm tries to propagate the value X through different branches around a fault candidate to the primary outputs, i.e., different sensitized paths are activated for each fault candidate. By activating different sensitized paths around a fault candidate, the intersection with paths related to other fault candidates is likely to be reduced, because the fault candidates are usually adjacent and close to each other in the circuit graph. As a result, the created counterexample usually decreases the number of fault candidates in the debugging procedure. In the following, first the branch and path activation method is described and then the complete algorithm for *Local Branch Activation* (LBA) is presented.

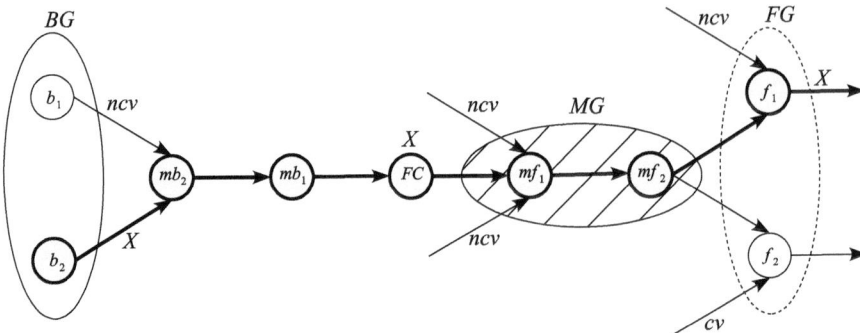

Fig. 3.3 Branch and path activation

3.3.1   Branch and Path Activation

A path is activated by propagating a value X through the corresponding path.
Conceptually, the value X is considered as an over-approximation for the behavior
of a fault candidate. Considering a given node FC, traversing the circuit graph
from node FC along successor nodes until reaching the first branch (fan-out), is
called *forward path* for FC. The successor nodes of the last node in a forward
path are called *forward nodes*. Traversing the circuit graph from node FC along
predecessor nodes until reaching the first branch (fan-in), is called *backward path*
for FC. The predecessor nodes of the last node in a backward path are called
backward nodes. Before reaching the forward nodes for FC, there are some middle
nodes which are collected in the set MG, and the forward nodes are collected in the
set FG. Figure 3.3 shows an example of a circuit graph. Considering FC, the sets
$MG = \{mf_1, mf_2\}$ and $FG = \{f_1, f_2\}$ result. Thus the number of forward nodes
(forward branches) is $|FG| = 2$. The backward nodes are collected in the set BG.
The set BG for FC in Fig. 3.3 is $BG = \{b_1, b_2\}$ and thus the number of backward
nodes (backward branches) is $|BG| = 2$. The X value of FC can be propagated
through $|BG| \times |FG|$ different paths. The method activates individual paths to more
likely reduce the intersection with paths related to other fault candidates. As shown
in Fig. 3.3, one of the paths is $P = (b_2, mb_2, mb_1, FC, mf_1, mf_2, f_1)$ indicated by
bold arrows. For propagating the X value of FC through the path P, the following
constraints are required (Fig. 3.3) using controlling and non-controlling values as
explained in Sect. 2.1 on Page 10:

– The off-path inputs of each gate $mf_i \in MG, i = 1, 2, \ldots, |MG|$, have to have
  a *ncv*.
– The off-path inputs of each gate $f_i \in FG, f_i \notin P, i = 1, 2, \ldots, |FG|$, have to
  have a *cv*.
– The off-path inputs of each gate $f_i \in FG, f_i \in P, i = 1, 2, \ldots, |FG|$, have to
  have a *ncv*.

Fig. 3.4 Automated
debugging using LBA method

```
1    algorithm  Debugging_LBA (In : Design, CE₁, Out : ℱ)
2    CEs = CE₁
3    New_CEs = ∅
4    do
5    {
6        CEs = CEs ∪ New_CEs
7        ℱ = SAT_Based_Debugging(Design, CEs)
8        foreach  Fault Candidate FC ∈ ℱ  do
9        {
10           Diag_Traces = LBA(Design, FC)
11           New_CEs = Testbench(Diag_Traces)
12           if  New_CEs! = ∅  then  break
13       }
14   }  while  New_CEs! = ∅
15   end  algorithm
```

- The output of each gate $b_i \in BG$, $b_i \notin P$, $i = 1, 2, \ldots, |BG|$, has to have a ncv.
- The output of each gate $b_i \in BG$, $b_i \in P$, $i = 1, 2, \ldots, |BG|$, has to have an X value.

3.3.2   LBA Algorithm

The algorithm of the complete procedure is listed in Fig. 3.4. A faulty design and an initial counterexample CE_1 are the input data of this algorithm (line 1). The initial counterexample is added to the set of counterexamples CEs (line 2). The first step is SAT-based debugging (line 7). SAT-based debugging finds and enumerates all fault candidates which can rectify the erroneous behavior of the current set of counterexamples (CEs). Then for each fault candidate FC, the function LBA generates some diagnostic traces (line 10). After that, the diagnostic traces are checked by the testbench to detect whether they lead to another counterexample (line 11). If at least one new counterexample is found, the algorithm continues the debugging step for the new set of counterexamples. The algorithm terminates when there is no new counterexample for any existing fault candidate.

Figure 3.5 shows the function LBA. This function generates diagnostic traces for a fault candidate. After converting the faulty design into CNF, additional constraints are inserted into the CNF. The constraint of line 2 assigns the value X to the output of a fault candidate. Line 3 causes that X to be observed at least at one PO. Then the algorithm searches for the backward and forward branches considering the fault candidate FC and constitutes different paths for X propagation (line 4) as explained in Sect. 3.3.1. For each path, the appropriate constraints as mentioned in the Sect. 3.3.1 are applied (line 8). Solutions are extracted and added to the set of diagnostic traces (lines 9–10).

In Fig. 3.5, there are three types of constraints: fault candidate constraint (line 2), observability constraint (line 3), and path constraint (line 8). For adding a new

Fig. 3.5 LBA function

```
1   algorithm  LBA (In : Design, FC, Out : Diag_Traces)
2   Add_Constraint(FC = X)
3   Add_Constraint(∑_{i=1}^{q}(PO_i == X) ≥ 1)
4   Paths = Find_Branches(FC)
5   Diag_Traces = ∅
6   foreach  Path ∈ Paths  do
7   {
8      Add_Constraint(Path)
9      if  Solve() == SAT  then
10        Diag_Traces = Diag_Traces ∪ Extract_Trace()
11  }
12  end  algorithm
```

constraint, firstly any previous constraint of the same type is removed, then the new constraint is inserted. For simplicity this is not shown in Fig. 3.5.

Each of the additional constraints increases the size of the SAT instance and consumes some memory which can be measured by the number of clauses. A fault candidate constraint requires one unit-literal clause. For each path, if standard gates are used in the circuit, two unit-literal clauses for backward nodes, m unit-literal clauses for middle nodes ($|MG| = m$), and two unit-literal clauses for forward nodes are added. For the observability constraint, $\mathcal{O}(q)$ clauses are inserted [SVAV05], where q is the number of primary outputs. Therefore, the number of additional clauses is $1 + (2 + m + 2) + \mathcal{O}(q) = \mathcal{O}(m + q)$.

Fault ranking techniques [JLJ09] may help increasing the performance of this algorithm. If a fault candidate related to the real bug location is processed earlier, then the diagnostic traces create at least one counterexample with higher probability.

BMC is utilized to create CNF for sequential circuits as explained in Sect. 2.2. For BMC, first the sequential circuit is unrolled for some time steps. In this case, each time step is like a combinational circuit and branches and paths are activated like in the combinational case. A fault candidate has one component in each time step. For applying the LBA method to sequential circuits, we constrain one fault candidate and its paths in each time step independently. Then the algorithm investigates whether the X can be observed on outputs. For multiple faults, where a fault candidate has multiple components, each component is activated independently. Therefore, the memory consumption for additional constraints remains as mentioned before.

Example 3.1. Consider the correct circuit of Fig. 3.6a which has been implemented faulty as Fig. 3.6b shows. An OR gate has been mistakenly implemented as an AND gate. There is an initial counterexample CE_1 which activates the bug in Fig. 3.6b. The input stimulus of CE_1 is $I_1 = (i_1, i_2, i_3, i_4, i_5) = (1, 0, 0, X, 1)$. The expected correct output response of CE_1 is $O_1 = (o_1, o_2) = (1, 1)$ while the faulty implementation produces $(X, 1)$. In the figure, the generated output o and its corresponding correct output o' are denoted by o/o'. Considering having CE_1, SAT-based debugging returns $\mathscr{F}_{CE_1} = \{b, c, d\}$ as fault candidates under the limitation $k = 1$. The outputs of the fault candidates of CE_1 are shown by black dots in the figure. Due to the limitation of $k = 1$, activating the abnormal predicate of either b, c, or d allows the corresponding gate to behave non-deterministically. In

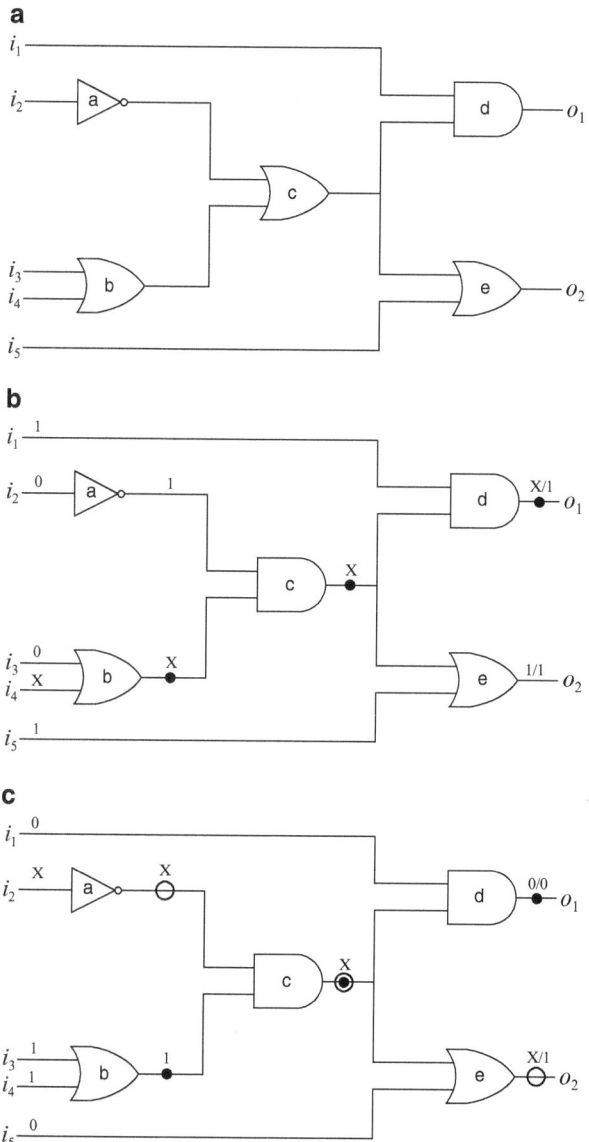

Fig. 3.6 LBA example. (**a**) Correct circuit. (**b**) Faulty implementation and initial counterexample CE_1. (**c**) New counterexample CE_2 generated by LBA

the figure, the correction blocks are not shown for simplicity. If value 1 is injected as correction value R_c through the correction block at the output of either b, c, or d, the erroneous behavior of CE_1 is fixed.

Now, LBA starts to activate the local branches of all fault candidates. One of the diagnostic traces leading to a counterexample is a diagnostic trace which is generated by activating the local branches of c. This diagnostic trace is shown in Fig. 3.6c which activates the path $P = (a, c, e)$. Therefore, the new counterexample CE_2 is generated with input stimulus $I_2 = (0, X, 1, 1, 0)$. The expected correct output response of CE_2 is $O_2 = (0, 1)$ while the faulty implementation produces $(0, X)$. The outputs of the fault candidates of CE_2 are shown by empty circles which are $\mathcal{F}_{CE_2} = \{a, c, e\}$. When both counterexamples CE_1 and CE_2 are used for debugging, the SAT solver searches for fault candidates which can fix the erroneous behavior of both counterexamples. Therefore, $\mathcal{F} = \{c\}$ is returned as the final set of fault candidates which precisely pinpoints the fault in the implementation.

3.4 Minimization of Sensitized Path Intersection

The *Minimization of Sensitized Path Intersection* (MSPI) heuristic finds the sensitized paths including a minimum number of existing fault candidates. Again, the don't care value X is used as a token for the behavior of one fault candidate or a set of fault candidates. The value X is propagated from inputs, crosses a number of fault candidates, and arrives at the outputs. The number of fault candidates having value X is denoted by L:

$$\sum_{i=1}^{|\mathcal{F}|} (FC_i == X) = L \tag{3.1}$$

The heuristic starts with $L = 1$ to find paths sensitizing one fault candidate. Thus, the value X is propagated from inputs, crosses one fault candidate, and arrives at outputs. If there is no path with one fault candidate having X which can create a diagnostic trace or a counterexample, L increases and the paths with more fault candidates having X are searched until at least one counterexample is found. The algorithm continues until L gets equal to the number of fault candidates.

The MSPI algorithm is shown in Fig. 3.7. The initial counterexample is assigned to the set of counterexamples CEs in Line 2. The set of new counterexamples New_CEs and L are initialized (lines 3–4). The first step is done by SAT-based debugging which extracts the fault candidates $\mathcal{F}$ (line 8). While L is less than the number of fault candidates and the set New_CEs is empty (line 10), the MSPI function searches for diagnostic traces (line 12). In the next step, the testbench checks the diagnostic traces for generating new counterexamples (line 13). In line 14, L is increased. If the diagnostic traces yield at least one new counterexample, the algorithm continues with the debugging step. Otherwise, the algorithm searches for new paths with more fault candidates having X. The algorithm finishes when L and the number of fault candidates converge.

The MSPI function is described in Fig. 3.8. First the faulty design is converted into CNF. Then additional constraints are inserted into the CNF. Line 2 fixes the number of fault candidates having an X value. This number is specified by L. Line 3 applies the observability constraint in order to observe the erroneous behavior at

Fig. 3.7 Automated debugging using MSPI method

```
1   algorithm  Debugging_MSPI (In : Design, CE₁, Out : ℱ)
2   CEs = CE₁
3   New_CEs = ∅
4   L = 1
5   do
6   {
7       CEs = CEs ∪ New_CEs
8       ℱ = SAT_Based_Debugging(Design, CEs)
9       New_CEs = ∅
10      while  L < |ℱ|  &&  New_CEs == ∅  do
11      {
12          Diag_Traces = MSPI(Design, ℱ, L)
13          New_CEs = Testbench(Diag_Traces)
14          L = L + 1
15      }
16  } while  New_CEs! = ∅
17  end algorithm
```

Fig. 3.8 MSPI function

```
1   algorithm  MSPI (In : Design, ℱ, L, Out : Diag_Traces)
2   Add_Constraint(∑(FCᵢ == X) = L)
3   Add_Constraint(∑(POᵢ == X) ≥ 1)
4   Diag_Traces = ∅
5   VisitedFCs = ∅
6   while  |VisitedFCs| < |ℱ|  do
7   {
8       NonVisitedFCs = ℱ \ VisitedFCs
9       Add_Constraint(∑(NonVisitedFCᵢ == X) ≥ 1)
10      if  Solve() == unSAT  then  break
11      Diag_Traces = Diag_Traces ∪ Extract_Trace()
12      VisitedFCs = VisitedFCs ∪ Extract_VisitedFCs()
13  }
14  end algorithm
```

least on one output. At this point, different methods can be applied for finding the diagnostic traces. One method can be simply finding all existing diagnostic traces according to the applied constraints (replacing lines 5–13). The weakness of this method is that usually there are many solutions with respect to L. Thus, the performance is biased by increasing the run time significantly. To overcome this weakness, the number of extracted diagnostic traces for each L should be limited by applying some heuristics. A heuristic method is applied in Lines 5–13. When a new diagnostic trace is found, the fault candidates having an X value on this diagnostic trace are added to the set of *VisitedFCs*. Next time, the algorithm searches for a diagnostic trace that includes at least one non-visited fault candidate (lines 8–9). By this, the diagnostic traces cover more fault candidates.

When the number of diagnostic traces in MSPI is limited, some useful counterexamples may not be found by the testbench. To overcome this weakness, another iteration is performed. When the MSPI algorithm terminates, there typically remains a small set of fault candidates. Executing MSPI for another round may find further counterexamples to reduce this set.

In Fig. 3.8, for applying the constraint related to the heuristic method (line 9) and the constraint on fault candidates (line 2), $\mathcal{O}(f)$ clauses are added to the CNF,

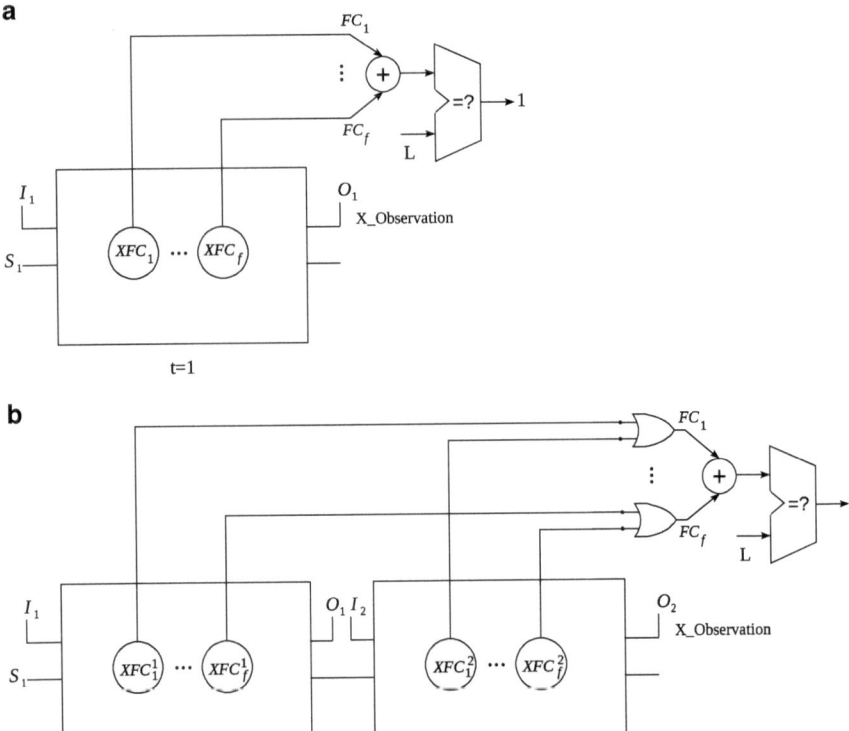

Fig. 3.9 MSPI method for sequential circuits with single faults: (**a**) first time step, (**b**) second time step

where f is the number of fault candidates. Also $\mathcal{O}(q)$ clauses are needed for the observability constraint on q primary outputs (line 3). Totally the algorithm requires $\mathcal{O}(f) + \mathcal{O}(q)$ additional clauses for combinational circuits in the case of a single fault.

The instance created by the MSPI method for single faults is shown in Fig. 3.9. When considering one time step for sequential circuits, one copy of the circuit is created (Fig. 3.9a). Then the X values of fault candidates are controlled by parameter L. When there are two time steps, two copies of the circuit are created. In this case, an OR gate controls the behavior of a fault candidate in different time steps (Fig. 3.9b).

In the case of multiple faults, each fault candidate $FC_i \in \mathscr{F}, i = 1, 2, \ldots, |\mathscr{F}|$, can have k components: $FC_i = \{FC_{i,1}, FC_{i,2}, \ldots, FC_{i,k}\}$. Also the erroneous behavior of a fault candidate can be propagated to outputs by each component or by a combination of components. This behavior is modeled by the following formula:

$$FC_i = \bigvee_{j=1}^{k} (FC_{i,j} == X) \tag{3.2}$$

Thus for each FC_i, the X behavior can be observed by assigning at least one of its components to X. For sequential circuits with a single fault, a similar strategy is applied. In this case each $FC_i \in \mathcal{F}, i = 1, 2, \ldots, |\mathcal{F}|$, has one component in each time step: $FC_i = \{FC_i^1, FC_i^2, \ldots, FC_i^s\}$, where s is the number of time steps. The activation of a fault candidate in each time or its activation by a combination of times may lead to the erroneous behavior on outputs. The following formula describes this behavior:

$$FC_i = \bigvee_{t=1}^{s} (FC_i^t == X) \tag{3.3}$$

For the case of sequential circuits with multiple faults, each fault candidate has two dimensions, one dimension represents the location and one dimension represents the time. In this case firstly in each time step, the sensitized components of a fault candidate are abstracted by Formula (3.4):

$$FC_i^t = \bigvee_{j=1}^{k} (FC_{i,j}^t == X) \tag{3.4}$$

Then the behavior of the fault candidate in time is abstracted according to Formula (3.5):

$$FC_i = \bigvee_{t=1}^{s} FC_i^t \tag{3.5}$$

Finally, for all of the above mentioned cases Formula (3.6) is used to apply the limitation (L) to all fault candidates:

$$\sum_{i=1}^{|\mathcal{F}|} FC_i = L \tag{3.6}$$

Using this method, the algorithm in Fig. 3.8 can be applied for all cases, i.e., single fault, multiple faults, combinational circuit and sequential circuit.

Figure 3.10 shows a sequential circuit with multiple faults in order to clarify the mentioned formulas. One time step is considered in Fig. 3.10a (this case is similar to combinational circuits with multiple faults). One OR-gate is inserted per fault candidate. The inputs of the OR-gates correspond to the variables of fault candidates specifying the X values. The outputs of the OR-gates are added and the sum is limited to L. Also the observability constraint is applied to the primary outputs.

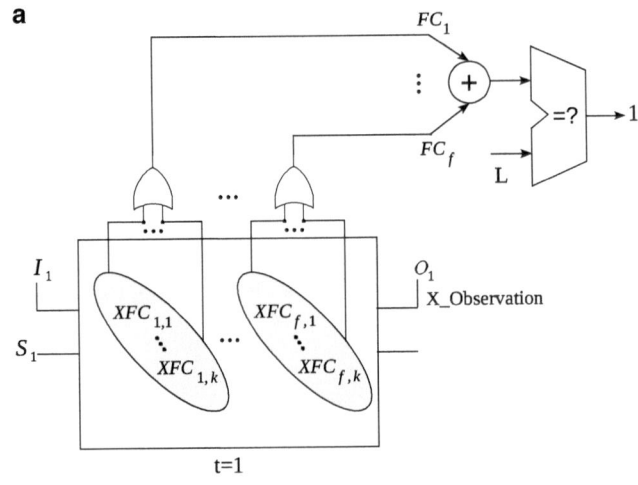

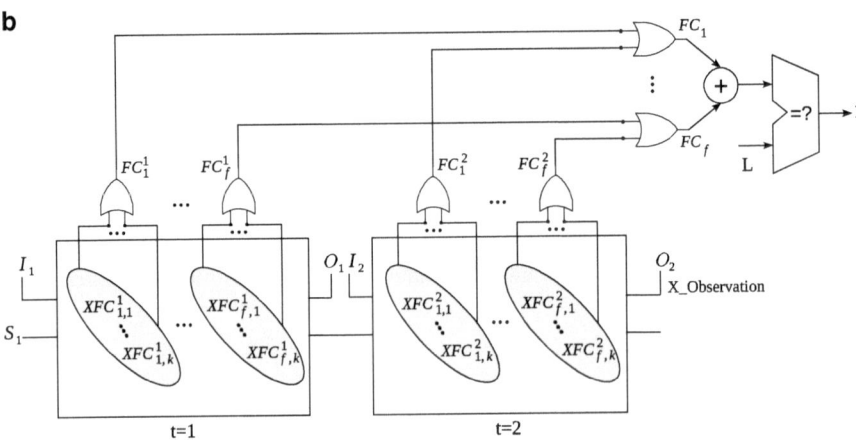

Fig. 3.10 MSPI method for sequential circuits with multiple faults. (**a**) First time step. (**b**) Second time step

Figure 3.10b shows two time steps. In addition to the OR-gates inserted for multiple faults, one OR-gate is applied for each fault candidate to control the fault candidate's behavior in the time dimension.

When considering $i \in [1, f]$ as fault candidate index, $j \in [1, k]$ as location index, and $t \in [1, s]$ as time index, the MSPI heuristic for sequential circuits with multiple faults needs $f \cdot s$ OR-gates with k inputs for each OR-gate ($f \cdot s \cdot OR(k)$) to control the locations of the fault candidates for all times. An OR gate with k inputs is denoted by $OR(k)$. Also, $f \cdot OR(s)$ are needed to control the time dimension. Thus, totally this method requires $\mathcal{O}(f) + \mathcal{O}(q) + f \cdot s \cdot OR(k) + f \cdot OR(s)$ additional constraints where each $OR(i)$ gate has $i + 1$ clauses. Therefore, the number of additional clauses is $\mathcal{O}(f) + \mathcal{O}(q) + f \cdot s \cdot (k + 1) + f \cdot (s + 1) = \mathcal{O}(f \cdot s \cdot k + q)$.

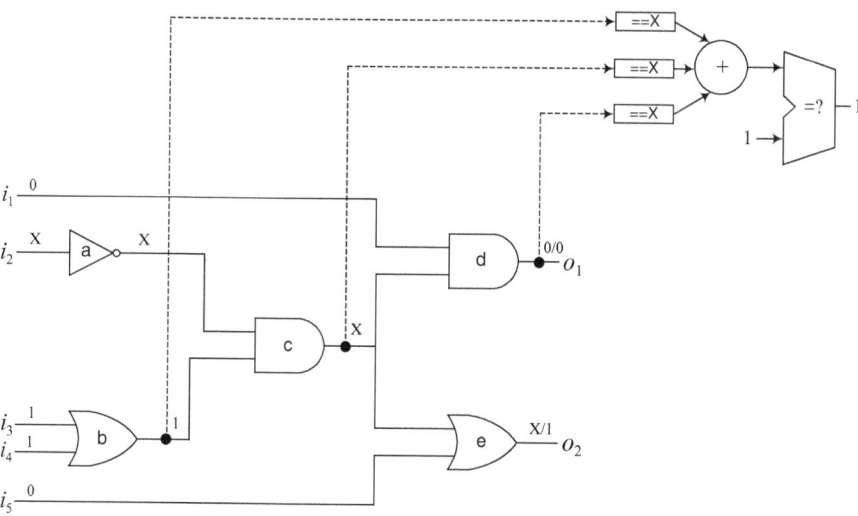

Fig. 3.11 MSPI example

Example 3.2. Again consider the faulty implementation and the initial counterexample CE_1 from Fig. 3.6b. After finding the fault candidates $\mathscr{F}_{CE_1} = \{b, c, d\}$ by SAT-based debugging, MSPI starts. MSPI adds the constraint of Formula (3.1) on fault candidates and starts with $L = 1$. This condition is shown in Fig. 3.11. After that, MSPI asks the SAT solver whether there is a diagnostic trace which sensitizes only one of the existing fault candidates. Therefore, the SAT solver searches for a sensitized path with one existing fault candidate having X.

Figure 3.11 shows one of the generated diagnostic traces which leads to a counterexample. This diagnostic trace has a sensitized path with minimum intersection to the existing fault candidates. Having the initial counterexample CE_1 and the new counterexample CE_2, the number of fault candidates effectively decreases to $\mathscr{F} = \{c\}$.

3.5   Limited Minimization Followed by Branch Activation

The counterexamples generated by LBA and MSPI may have different characteristics. LBA investigates each fault candidate in detail whereas MSPI has an overall view of all fault candidates. We propose one unified algorithm called *Limited Minimization followed by Branch Activation* (LMBA) to combine the advantages of both methods and, consequently, to obtain higher accuracy. The LMBA heuristic tries to increase the diagnosis accuracy with a reasonable overhead by improving the MSPI heuristic with local branch activation. After finishing the MSPI algorithm with a small set of fault candidates, activating the branches of all fault candidates

Fig. 3.12 Automated debugging using LMBA method

```
1   algorithm  Debugging_LMBA (In : Design, CE₁, Out : ℱ)
2   CEs = CE₁
3   New_CEs = ∅
4   L = 1
5   do
6   {
7       CEs = CEs ∪ New_CEs
8       ℱ = SAT_Based_Debugging(Design, CEs)
9       New_CEs = ∅
10      while  L < |ℱ|  &&  New_CEs == ∅  do
11      {
12          Diag_Traces = MSPI(Design, ℱ, L)
13          New_CEs = Testbench(Diag_Traces)
14          L = L + 1
15      }
16
17      if  L == |ℱ|  then
18      {
19          foreach  Fault  Candidate  FC ∈ ℱ  do
20          {
21              Diag_Traces = LBA(Design, FC)
22              New_CEs = New_CEs ∪ Testbench(Diag_Traces)
23          }
24          L = L + 1
25      }
26
27  }  while  New_CEs! = ∅
28  end  algorithm
```

can be done in a short time and without high computational cost. By spending this short time, a higher accuracy can likely be achieved.

The details of the LMBA heuristic are shown in Fig. 3.12. First the algorithm searches for the paths including a minimum number of Xs on the fault candidates (lines 10–15). When the convergence of L and the number of fault candidates is reached, the first step of LMBA is finished. Thus, the second step of LMBA starts. Now, the local branches of all fault candidates are activated (line 19–23) and the new counterexamples are collected (line 22). If there is at least one new counterexample, then SAT-based debugging is executed. After that the LMBA algorithm finishes. The number of required additional clauses for the LMBA method is the maximum of additional clauses of LBA and MSPI: $\max(\mathcal{O}(m + q), \mathcal{O}(f \cdot s \cdot k + q))$.

3.6 Experimental Results

The effects of the proposed heuristic techniques are experimentally demonstrated in this section. The proposed techniques of this chapter are implemented using C++ in the WoLFram environment [SKF⁺09] and are evaluated on combinational and sequential circuits of LGSynth-93 [McE93] and ITC-99 [Dav99] benchmark suites.

The faults are randomly injected by replacing gates. For example an AND gate is replaced by an OR gate. The circuits are unrolled for five time steps for bounded sequential debugging.

The experiments are carried out on a Quad-Core AMD Phenom(tm) II X4 965 Processor (3.4 GHz, 8 GB main memory) running Linux. MiniSAT is used as underlying SAT solver [ES04]. Run time is measured in CPU seconds, and the memory consumption is measured in MB.

We compare the methods proposed in this chapter (LBA, MSPI, LMBA) to a method based on random trace generation (RND) and to QBF [SFD10]. Note that LBA, MSPI and LMBA do not have access to a formal specification but only use a testbench as a simulation model or a black box specification. Thus, the accuracy of the proposed methods is limited. In contrast, QBF [SFD10] uses a formal specification and can therefore achieve a higher accuracy. However, a formal specification is often not given for complex designs. In these experiments all methods are limited to a maximum of ten iterations between the debugging and the verification procedures. As mentioned in Sect. 3.4, the MSPI method may be executed for two rounds in these ten iterations. The bold numbers in the tables show the best results among heuristics and random methods.

The experimental results for single faults are presented in Tables 3.1 and 3.2. Table 3.1 shows the final number of fault candidates (#FC) and the total number

Table 3.1 Results for single faults (diagnosis accuracy)

Method		Heuristic Methods						Rand.		Form.	
Name		LBA		MSPI		LMBA		RND		QBF	
Circuit	#Gates	#FC	#CE	#FC	#CE	#FC	#CE	#FC	#CE	#FC	#CE
comb.											
apex5	3938	**2**	7	**2**	7	**2**	7	3	21	2	4
c7552	4674	22	18	22	17	22	18	22	21	22	6
cordic	2938	10	18	10	8	10	12	10	21	10	3
dalu	2883	4	15	4	9	4	13	4	21	4	2
des	3942	2	3	2	3	2	3	2	21	2	2
i10	3294	**6**	18	12	11	7	17	14	21	6	3
misex3	6249	**2**	15	**2**	7	**2**	7	4	21	2	3
pair	2848	**9**	11	**9**	10	**9**	14	10	21	9	3
seq	4776	**7**	14	**7**	5	**7**	9	9	21	7	3
seq.											
b04	821	**6**	20	15	20	15	21	18	21	6	7
b05	1198	2	1	2	1	2	1	2	21	2	1
b08	223	6	1	6	1	6	1	6	21	4	2
b10	260	**1**	21	11	14	10	16	10	21	1	3
b11	867	9	21	9	3	9	8	9	21	5	3
b12	1297	29	6	**27**	21	**27**	20	29	21	27	3
gcd	1217	7	19	7	21	7	21	7	21	7	4
phase_de.	1834	29	3	29	16	29	20	29	21	29	2

Table 3.2 Results for single faults (time and memory)

Method		Heuristic Methods						Rand.		Form.	
Name		LBA		MSPI		LMBA		RND		QBF	
Circuit	#Gates	Time	Mem	Time	Mem	Time	Mem	Time	Mem	Time	Mem
comb.											
apex5	3938	19.8	39	12.7	35	13.4	35	66.4	104	22.8	26
c7552	4674	639.9	188	712.8	204	538.7	139	691.7	206	346.5	77
cordic	2938	109.5	78	46.1	39	43.7	51	260.7	102	30.6	19
dalu	2883	82.1	102	14.1	39	29.7	52	49.2	104	14.9	19
des	3942	3.7	18	6.6	26	6.5	26	70.8	138	11.1	18
i10	3294	886.5	104	253.8	102	360.7	103	593.2	137	52.5	20
misex3	6249	126.9	105	29.1	52	24.5	52	443.9	140	18.8	38
pair	2848	182.5	69	87.2	68	84.7	69	192.3	96	41.3	26
seq	4776	141.4	103	39.1	38	46	52	176.6	138	19.7	34
seq.											
b04	821	115.5	105	158.6	106	146.1	106	134.8	106	57.8	47
b05	1198	0.3	9	0.3	9	0.3	9	26.1	140	4.9	25
b08	223	0.6	2	0.3	2	1.1	2	6.8	26	1.1	5
b10	260	26.6	51	17.3	26	18.2	26	17.6	39	2.2	8
b11	867	73.6	138	7.6	23	21.2	47	71.3	138	10.8	34
b12	1297	69.1	68	329.2	207	201.4	189	328.1	207	78.1	51
gcd	1217	150.2	188	151.8	205	128.2	157	135.7	189	81.4	50
phase_de.	1834	45.1	48	388.5	209	434.2	277	471.4	278	113.5	47

of counterexamples used (#CE). Table 3.2 shows the required run time (*Time*), and the maximum memory consumption (*Mem*). The experiments are started with one initial counterexample obtained from a verification tool. The QBF method [SFD10] is the only exact approach in the sense that it takes all counterexamples into account. Thus, the number of fault candidates (#FC) determined by the QBF method is considered as the minimal number of fault candidates that may be determined. In the tables, the sign > 99 indicates more than 99 fault candidates. The experiments of Table 3.1 show that the diagnosis accuracy of the presented methods is better than of RND. By comparing #FC columns, LBA, MSPI and LMBA have a better accuracy than RND in 7 experiments, while RND is never better than LBA or LMBA. Also for single faults, LBA has the best accuracy among the methods presented in this chapter in most of the experiments. LBA is as accurate as QBF [SFD10] in 82 % of the experiments (14 experiments), while MSPI and LMBA are as accurate as QBF [SFD10] in 71 % of the experiments (12 experiments). Overall, Fig. 3.13 compares the heuristic and random methods to the exact formal method with respect to the diagnosis accuracy. The diagram shows for which percentage of the experiments, each heuristic method is as accurate as the exact formal method.

For $i10$ in Tables 3.1 and 3.2, LBA has a good accuracy but long run time. LMBA obtains a good accuracy with a reasonable run time by merging the advantages of

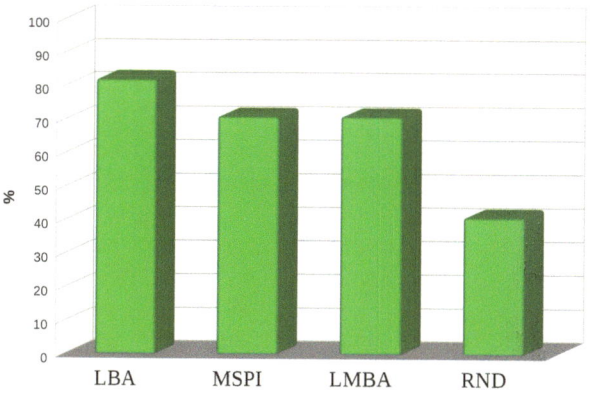

Fig. 3.13 Diagnosis accuracy of the methods for single faults

Table 3.3 Results for multiple faults (diagnosis accuracy)

Method		Heuristic Methods									Rand.			Form.		
Name		LBA			MSPI			LMBA			RND			QBF		
Circuit	#Gates	k	#FC	#CE	k	#FC	#CE	k	#FC	#CE	k	#FC	#CE	k	#FC	#CE
comb.																
apex5	3938	1	5	14	3	>99	19	3	>99	19	2	45	21	3	60	5
c7552	4674	1	4	21	2	88	21	1	2	4	1	2	21	2	58	9
cordic	2938	1	9	21	1	6	11	1	6	12	1	9	21	1	3	3
dalu	2883	**2**	**56**	16	**2**	**56**	19	**2**	**56**	18	2	>99	21	2	56	7
des	3942	3	4	20	3	4	12	3	4	13	3	4	21	3	4	3
i10	3294	1	2	19	1	1	6	1	1	7	1	1	21	2	6	6
misex3	6249	2	26	21	**2**	**16**	20	2	18	21	2	60	21	2	16	5
pair	2848	2	15	20	2	15	11	**3**	>99	12	2	15	21	3	>99	6
seq	4776	2	4	21	**2**	**2**	11	**2**	**2**	21	1	3	21	2	2	5
seq.																
b04	821	1	5	21	1	2	6	1	2	14	1	2	21	2	62	13
b05	1198	1	5	1	1	5	5	1	5	3	1	5	21	1	5	1
b08	223	1	69	1	**1**	**58**	20	**1**	**58**	20	**1**	**58**	21	1	58	2
b10	260	2	14	11	2	14	17	2	14	8	2	14	21	2	14	6
b11	867	**2**	**60**	11	2	66	1	2	66	1	**2**	**60**	21	2	60	3
b12	1297	1	93	1	1	93	12	1	93	12	1	93	21	1	93	1
gcd	1217	2	35	12	**2**	**28**	19	**2**	**28**	17	2	32	21	2	28	3
phase_de.	1834	1	11	13	1	11	19	1	11	18	1	11	21	1	11	2

LBA and MSPI. Also LMBA and MSPI are faster than LBA for combinational circuits. In sequential circuits $b04$ and $b10$, LBA finds the minimum number of fault candidates in a relatively short time.

Tables 3.3 and 3.4 show the experimental results for multiple faults. In this case, each fault candidate has up to k components. The debugging procedure starts with

Table 3.4 Results for multiple faults (time and memory)

Method		Heuristics Methods						Rand.		Form.	
Name		LBA		MSPI		LMBA		RND		QBF	
Circuit	#Gates	Time	Mem	Time	Mem	Time	Mem	Time	Mem	Time	Mem
comb.											
apex5	3938	76.5	102	1447.4	153	1444.7	153	250.1	154	1107.6	124
c7552	4674	260.2	189	1846.9	303	41.2	47	143.8	189	877.1	185
cordic	2938	77.4	102	24.7	51	22.1	39	65.4	102	7.6	19
dalu	2883	395.7	184	424.1	185	561.5	184	1270.1	379	194.4	124
des	3942	256.6	139	85.3	78	86.5	78	174.9	140	205.3	74
i10	3294	101.3	104	23.4	47	31.8	39	65.8	105	84.5	68
misex3	6249	586.1	157	440.4	123	221.1	123	3270.2	254	145.3	76
pair	2848	169.4	136	94.2	69	193.8	160	151.7	136	2453.1	158
seq	4776	289.5	186	92.2	93	292.4	121	98.6	137	34.4	67
seq.											
b04	821	116.4	106	20.3	51	36.1	70	67.3	106	1082.6	152
b05	1198	4.2	9	20.4	47	11.8	26	41.9	140	11.2	23
b08	223	6.2	4	61.6	26	62.4	26	66.1	38	4.5	4
b10	260	222.7	62	120.1	62	61.4	19	99.5	59	16.3	43
b11	867	1053.7	215	32.1	47	87.6	47	781.8	251	223.2	467
b12	1297	41.9	19	269.2	105	269.1	105	787.7	207	171.2	33
gcd	1217	248.4	153	563.3	189	342.2	139	332.6	206	40163.1	141
phase_de.	1834	215.5	207	374.1	276	298.1	211	303.6	278	41.9	51

$k = 1$ and iteratively increases k until a satisfying solution is found. This yields a fault candidate which is a tuple of k components. Due to fault model-free SAT-based debugging, fault masking may not be recognized here (as mentioned in Sect. 2.4.1).

The methods which do not have access to a formal specification have some limitations for their diagnosis accuracy. One limitation occurs when one fault (among multiple faults) always remains inactive and never appears among the fault candidates found by the counterexamples so far. This can be seen in Table 3.3 for $i10$. MSPI and LMBA localize one fault location accurately but another fault remains always inactive ($k = 1$), while QBF [SFD10] activates all faults ($k = 2$) by comparing the faulty circuit to a complete specification.

When comparing two approaches, a larger value of k indicates better accuracy. If the value of k is equal for two approaches, a smaller #FC indicates better accuracy. In 7 experiments, MSPI and LMBA have a better accuracy than RND. MSPI and LMBA are less accurate than RND in only one experiment, while in that case they use less counterexamples. MSPI and LMBA are as accurate as QBF [SFD10] in 59 % of the experiments (10 experiments) for multiple faults. Figure 3.14 presents the overall comparison of heuristic and random methods to the exact formal method for multiple faults.

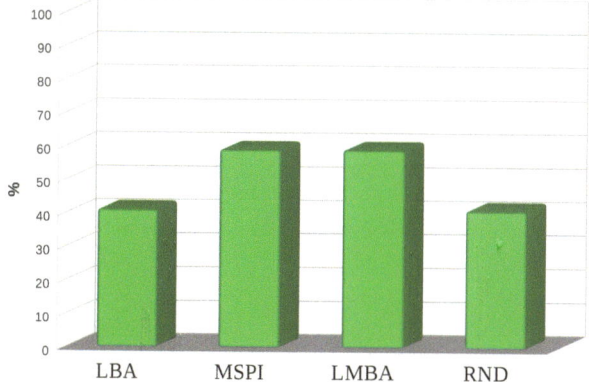

Fig. 3.14 Diagnosis accuracy of the methods for multiple faults

For *c7552* in Table 3.3, MSPI determines a minimum cardinality of 2 which is found by QBF [SFD10], too. For *misex3*, MSPI finds the minimum number of fault candidates. LMBA also has a good accuracy with a reasonable time for multiple faults. For *pair*, LMBA determines a minimum cardinality of 3 while this method does not spend a long time. Also for *gcd*, LMBA has the best performance and accuracy.

Although the QBF [SFD10] method as expected has the best accuracy in the experiments, the heuristic methods often have a better performance for complex circuit structures and for fault candidates of large cardinality. For example for *pair* and *gcd* in Tables 3.3 and 3.4, the QBF [SFD10] approach is slow, while the heuristic approaches have a good performance and accuracy. This is due to the fact that the heuristic methods do not need to enumerate fault candidates explicitly. Especially, an explicit enumeration for multiple faults reduces the performance of the exact approach, because the number of fault candidates increases exponentially with the fault cardinality [SFD10].

The experimental results show when there is no formal specification, LBA is a good method for the diagnosis of single faults while MSPI and LMBA are suitable for multiple faults. Thus, by selecting the best heuristics our flow is as accurate as an exact formal debugging in 71 % of the 24 experiments. The time spent for debugging and verification for single fault experiments is shown in Table 3.5. The methods presented in this chapter usually spend a shorter time for the verification while the debugging time is longer because the number of counterexamples may be larger.

Table 3.5 Debugging and verification times for single fault experiments

Method		LBA		MSPI		LMBA		RND		QBF	
Circuit	#Gates	Time (s)		Time (s)		Time (s)		Time (s)		Time (s)	
comb.		Deb.	Ver.	Deb.	Ver.	Deb.	Ver.	Deb.	Ver.	Deb.	Ver.
apex5	3938	14.83	5.06	7.26	5.45	7.79	5.7	63.63	2.77	10.28	12.59
c7552	4674	515.01	124.94	623.88	88.99	353.82	184.89	674.78	17	154.03	192.51
cordic	2938	80.27	29.28	21.53	24.6	19.02	24.68	68.8	191.92	12	18.66
dalu	2883	31.97	50.22	9.48	4.61	11.81	17.91	44.16	5.04	9.29	5.7
des	3942	1.96	1.81	3.11	3.52	2.96	3.6	66.16	4.72	2.31	8.79
i10	3294	255.94	630.58	124.86	128.98	183.34	177.36	267.64	325.62	22.25	30.3
misex3	6249	113.57	13.36	22.85	6.17	18.19	6.34	440.48	3.47	7.57	11.26
pair	2848	144.23	38.36	49.64	37.64	41.45	43.28	135.05	57.33	20.87	20.52
seq	4776	114.18	27.29	15.09	23.95	15.55	30.45	129.02	47.58	7.19	12.56
seq.											
b04	821	94.76	20.76	131.97	26.68	118.15	27.99	128.89	5.96	5.31	52.51
b05	1198	0.36	0.02	0.35	0.02	0.36	0.02	19.62	6.55	0.74	4.22
b08	223	0.11	0.49	0.1	0.27	0.11	0.93	6.36	0.51	0.13	1.02
b10	260	21.14	5.47	11.62	5.69	9.97	8.23	16.5	1.15	0.23	2.01
b11	867	51.74	21.93	3.14	4.54	5.2	16.09	67.37	3.98	1.68	9.17
b12	1297	45.18	23.98	285.45	43.8	152.86	48.59	320.31	7.85	6.71	71.3
gcd	1217	120.76	29.47	112.93	38.88	88.52	39.77	125.58	10.14	3.94	77.5
phase de.	1834	23.39	21.69	311.17	77.34	323.44	110.85	458.12	13.31	8.98	104.52

3.7 Summary

This chapter proposed an approach for automating the design debugging procedures by integrating SAT-based debugging with testbench-based verification. The diagnosis accuracy increases by iterating debugging and counterexample generation, i.e., the total number of fault candidates decreases.

Three techniques were proposed to generate diagnostic traces for deriving high quality counterexamples enhancing the diagnosis accuracy. LBA activates the local branches of each fault candidate. MSPI finds the sensitized paths including a minimum number of fault candidates. The advantages of both techniques are combined in LMBA. These techniques were evaluated and compared to random trace generation and one completely formal technique with respect to accuracy, run time and memory. The experimental results showed that our approach has an accuracy close to an exact formal debugging approach.

This chapter focused on the design debugging at the pre-silicon stage. In the next chapter, we exploit the approach of this chapter in order to develop a generalized debugging framework which can be utilized in both pre-silicon and post-silicon stages.

Chapter 4
Automated Debugging from Pre-Silicon to Post-Silicon

Different approaches have been proposed for automating pre-silicon and post-silicon debugging. Automated approaches in pre-silicon debugging rely on simulation [VH99], BDD [CH97], and SAT [SVAV05]. In [SD10], SAT-based debugging is used to debug different abstraction levels of the system description. Post-silicon debugging requires a larger effort. The post-silicon validation process starts by applying test vectors to the IC or by running a test program, such as end-user applications or functional tests, on the IC until an error is detected [CMB07a, PHM09]. The erroneous behavior and golden responses obtained by system simulation constitute a counterexample. Having a counterexample, post-silicon debugging is carried out to localize and rectify the root cause of the erroneous behavior.

One main challenge of post-silicon debugging is the limited observability of internal signals. To address this problem, various on-chip solutions for internal signal observation have been proposed including ones based on *Design-For-Test* (DFT) structures such as scan chains [HMM06, VWB02] and based on *Design-For-Debug* (DFD) structures such as trace buffers [ABD+06, YT08, LMF11]. The techniques based on trace buffers store internal signal traces in on-chip memories and are widely accepted in industry [LMF11].

Some approaches have been proposed for automating post-silicon debugging. The work in [PHM09] uses trace buffer data with self-consistency-based program analysis techniques for bug localization. In [YNV09], trace buffer data is analyzed to detect errors in both the spatial and the temporal domains. The analysis provides suggestions for the setup of the test environment in the next debug session by giving a better estimate for the window (time interval) of cycles the engineer should concentrate on catching the error. Even by using the trace buffers, getting an execution trace of the on-chip signals related to the time of bug activation is a challenging problem. To address this problem, different methods for bugs and faults have been proposed. In [dPNN+11], the approach re-runs the chip with new trigger conditions to "backspace" the content of the trace buffer until the traces related to

© Springer International Publishing Switzerland 2015
M. Dehbashi, G. Fey, *Debug Automation from Pre-Silicon to Post-Silicon*,
DOI 10.1007/978-3-319-09309-3_4

the activation time of the design bug can be extracted. In [THPM$^+$10] and [PHM09], some quick error detection mechanisms for electrical faults are used to efficiently store the erroneous behavior related to the bug activation time in the trace buffer.

In this chapter, a generalized framework to automate debugging is proposed [DF12] that tightly integrates MBD [Rei87] using SAT [SVAV05] and diagnostic trace generation [DSF11]. The main contributions of this chapter are:

- a unified view on pre- and post-silicon debugging automation and
- a detailed discussion of post-silicon debugging scenario

Our approach relies on model-based diagnosis as an underlying step. Diagnostic traces [DSF11] close the loop between verification and model-based diagnosis. Diagnostic traces differentiate the fault candidates and increase the diagnosis accuracy. The debugging flow can be applied to electrical faults as well as design bugs which slip into the IC from different levels of the system description. An instantiation of the generalized automated debugging flow is applied to post-silicon debugging of design bugs.

At post-silicon stage, we use trace buffers as a hardware structure for debugging. Post-silicon debugging is automated by integrating post-silicon trace analysis, model-based diagnosis [SVAV05], and diagnostic trace generation [DSF11]. The flow closes the loop between post-silicon verification and debugging which increases the diagnosis accuracy and decreases the debugging time. As a result, the time of the IC development cycle decreases significantly and the productivity increases.

In the flow, a designer investigates the sensitized paths leading to erroneous behavior on a schematic view of the circuit. The sensitized paths leading to erroneous behavior are highlighted by the fault candidates discovered by model-based diagnosis. Then, diagnostic trace generation tries to create more counterexamples which help the designer explaining erroneous behavior with fewer fault candidates. Thus, the designer can focus on a small section of the circuit to do the final rectification spending a short time.

The remainder of this chapter is organized as follows. Section 4.1 introduces preliminary information on hardware structures for post-silicon debugging. Then, our generalized approach to automate debugging is presented in Sect. 4.2. Section 4.3 describes our automated debugging approach for post-silicon to diagnose bugs and faults. A concrete instantiation of our approach for design bugs is presented in Sect. 4.4. Section 4.5 presents experimental results on benchmark circuits. Section 4.6 summarizes this chapter.

4.1 Hardware Structures for Post-Silicon Debugging

Hardware structures for post-silicon debugging are divided into two main categories: DFT structures and DFD structures. Scan chains are commonly used as a DFT structure in manufacturing test. This hardware can be reused for post-silicon

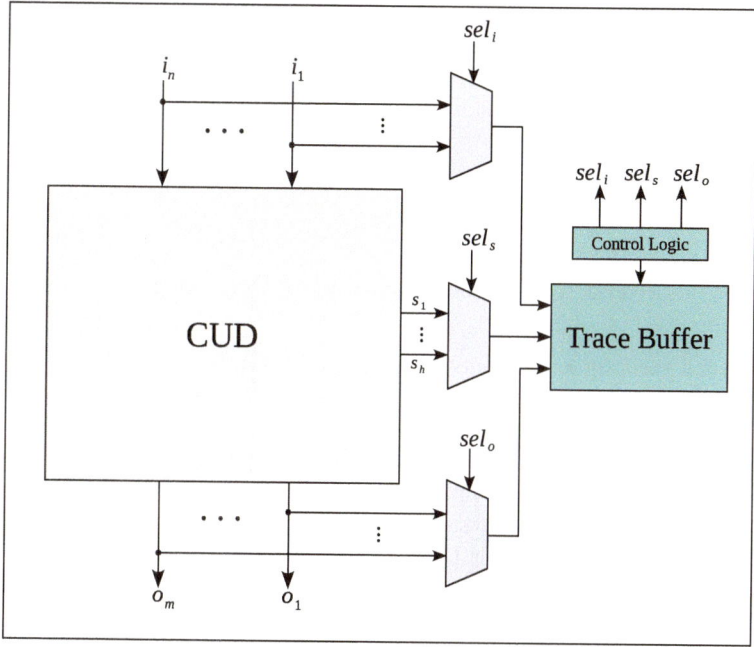

Fig. 4.1 Hardware structure

debugging [VWG02]. During the test mode, the state of all the scan registers can be extracted by performing a scan dump. Unless two-state elements are used for each register, which leads to an excessive area overhead, the test environment needs to be restarted after each scan dump [YNV09]. The scan registers with two-state elements are used in [CMA08] for online detection of design bugs.

An overview of DFD structures is given in [ABD+06]. In industry, trace buffers are commonly utilized as a DFD structure. The hardware structure of a trace buffer is shown in Fig. 4.1. Control logic is responsible to detect trigger conditions so it can determine when and which data signals will be sampled in the trace buffer. The trigger conditions can be based on time, event, or a mixture of both. The control logic tries to sample signals in a way that the best verification coverage is obtained. For verifying the trace buffer data after running a test program, some inputs, outputs and internal states of the circuit need to be stored in the trace buffer. The signals that need to be sampled at the same time should have independent entries to the trace buffer.

The control logic may divide the trace buffer into multiple segments [AN07]. These segments may belong to signals within different entities or signals related to different events. Also segmentation may be based on the sampling period [YNV09]. For example a trace buffer can have two segments, the first segment for samples from clock cycle 100 to 500, the second segment for samples from clock cycle 800 to 1200. Also the control logic can control the trace buffer as a circular buffer

[PHM09]. In this case, information related to the last events can be recovered from the trace buffer. After finishing the test program or detecting an error, the control logic serializes the content of the trace buffer and sends it back to the off-chip debugger software via a low-bandwidth interface such as JTAG [AN07]. This data constitutes the initial counterexamples. The trace buffer size in practice is typically $1 \text{ K} \times 8$ bits to $8 \text{ K} \times 32$ bits [KN08, YT13].

4.2 Generalized Automated Debugging Procedure

We present a generalized *Automated Debugging* (AD) procedure that utilizes MBD and diagnostic traces for automation and accurate localization of potential root causes of an error. The function AD was inspired from the debug flow presented in the previous chapter. AD can be reused in different contexts for various debugging situations. Diagnostic traces help automated debugging by distinguishing fault candidates. Actually, a concept similar to diagnostic traces is used not only for debugging in computer science but also in other sciences (e.g. medical science, psychology, ...) to accurately discover disorders in clinical cases (as fault candidates) out of a community (as a system) [KM03].

The diagnosis in a system begins with system observations violating normal system behavior determined by the system specification. This discrepancy between a system and its specification is called counterexample. Having an initial counterexample, MBD tries firstly to localize fault candidates. Fault candidates are components of the system capable of rectifying the erroneous behavior. But usually the initial number of fault candidates is large. Thus, diagnostic traces are introduced to discriminate the fault candidates and to help debugging to accurately localize the root cause of the error. Diagnostic traces discover more erroneous behavior. This new behavior should be used to iterate MBD for excluding fault candidates that cannot fix the new erroneous behavior.

The inputs of the AD function are a system specification as a reference (*Ref*), a system model as an object (*Obj*), and one or more initial counterexamples (*CEs*) which show the initial discrepancy between the system and its specification. The output of AD is the set $\mathcal{F}$ of fault candidates which are the potential sources of the observed error corresponding to the available counterexamples. Each fault candidate is a set of components of the system which can fix all erroneous behavior of the counterexamples under consideration:

$$\mathcal{F} = AD(Ref, \ Obj, \ CEs) \tag{4.1}$$

The AD function consists of three subfunctions. The first subfunction of AD applies MBD. MBD finds the initial set $\mathcal{F}'$ of fault candidates as potential sources of the observed errors according to the initial counterexamples:

$$\mathcal{F}' = MBD(Obj, \ CEs) \tag{4.2}$$

Diagnosis accuracy is a function of fault candidates, e.g., a small number indicates good accuracy. In general, the value of the diagnosis accuracy is determined by the quantity and quality of fault candidates. If the diagnosis accuracy determined by fault candidates is not sufficient, then the second step of AD starts. The second step is *Diagnostic Trace Generation* (DTG). Similar to diagnostic test patterns [ZCY$^+$07, GMK91], diagnostic traces distinguish the behavior of fault candidates [DSF11]. DTG works on sequential circuits and does not require a precise fault model. Diagnostic traces may help the next debugging session to exclude fault candidates which cannot fix all erroneous behavior:

$$DTs = DTG(Obj, \mathscr{F}') \qquad (4.3)$$

Afterwards, diagnostic traces are validated with respect to the system specification to guarantee that the diagnostic traces really create a discrepancy between the system and its specification. A diagnostic trace which creates a discrepancy is a counterexample. This step is called *Diagnostic Trace Validation* (DTV):

$$CEs = DTV(Ref, DTs) \qquad (4.4)$$

The debug process is iterated by using the new counterexamples to decrease the number of fault candidates. The automated debugging function is shown in Fig. 4.2 as a pseudo code. AD is controlled by a parameter for the threshold of the *Diagnosis Accuracy* (line 6), e.g., the number of fault candidates can be used as a value for the threshold parameter. The number of diagnostic traces is controlled by an expression *DTsLimitation* which compares the total number of generated diagnostic traces with a limit (line 8).

To apply the generalized approach to pre-silicon debugging, *Obj* is a hardware design at the RTL or the gate level. *Ref* can be a formal specification or a testbench. Initial counterexamples *CEs* are given by design verification tools. In this case, AD is invoked to search the set $\mathscr{F}$ of fault candidates such that the appropriate diagnosis accuracy is achieved [DSF11]:

```
1   algorithm AD (In : Ref,Obj,CEs, Out : 𝓕)
2   do
3   {
4       𝓕' = MBD(Obj,CEs)
5       NewCEs = ∅
6       if DiagnosisAccuracy(𝓕') < Threshold then
7       {
8           while NewCEs == ∅ && !DTsLimitation do
9           {
10              DTs = DTG(Obj, 𝓕')
11              NewCEs = DTV(Ref, DTs)
12          }
13          CEs = CEs∪NewCEs
14      }
15  } while NewCEs! = ∅
16  𝓕 = 𝓕'
17  end algorithm
```

Fig. 4.2 Generalized automated debugging procedure

$$\mathscr{F} = AD(Spec, \ Design, \ CEs) \tag{4.5}$$

An instantiation of the function AD for pre-silicon debugging leads to the debug automation flow proposed in Chap. 3.

The work in [SFD10] integrates the functions *DTG* and *DTV* in a unified instance to generate new counterexamples. But in that work, *Ref* is a formal specification. However, a formal specification is often not given for complex designs. In the following, we explain how AD is utilized in post-silicon debugging.

4.3   Automated Post-Silicon Debugging

In this section, we explain how the general automated debugging procedure is used for post-silicon debugging of design bugs and electrical faults. In the IC design hierarchy, there can be multiple references for a chip as a system. By using different references, bugs and faults are distinguished. As shown in Fig. 4.3, for post-silicon debugging usually there are three main descriptions of the hardware system: specification, design, and chip.

The specification as a golden model can be a formal specification, a high level simulation model or a testbench. The specification is used for creating the expected correct output of a trace in the debugging process. A design is a circuit which is represented at RTL using hardware description languages. Then, logic synthesis and place-and-route processes create the gate level and the transistor level designs respectively for chip manufacturing.

After manufacturing the chip, post-silicon validation begins by running a test program, such as an end-user application or functional tests, or applying the test vectors. Hardware or software assertions may observe an error. In this case, signal traces are stored in trace buffers. The content of the trace buffer is used to extract the system state, its inputs, and the corresponding outputs. Then, the specification and the design are used to check the extracted traces. There are some shared observation points among specification, design, and chip. This yields knowledge about equivalent states in the different abstraction levels which can be utilized for debugging.

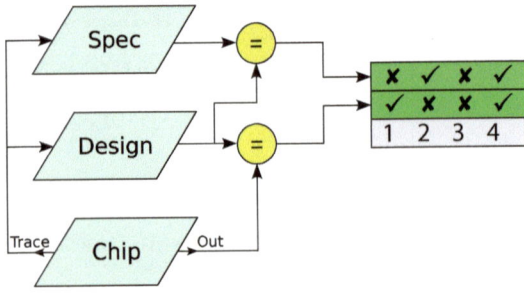

Fig. 4.3 The start of debugging after an error detection on a chip. ✗ shows the inconsistency between the corresponding responses in each level

Our general idea is based on inconsistencies detected between different system descriptions which can be used to distinguish bugs and faults. As Fig. 4.3 shows, after an error occurrence, the traces and their corresponding responses are extracted from the trace buffer. Checking the extracted traces from the trace buffer in the specification and in the design leads to four cases. For each case, the general AD function is configured in different ways to efficiently diagnose bugs. Each case is discussed in the following.

4.3.1   Case 1: Design Bug

In this case (Fig. 4.3, case 1), the extracted traces applied to the design create responses which are consistent with the extracted responses from the trace buffer. When the traces are applied to the specification and the design, different responses are created. This indicates there is a bug in the design which has escaped pre-silicon verification and has slipped into the chip. Actually, the extracted traces show a special sequence of the system traces which was not verified in the pre-silicon verification. This is revealed after running the chip for a longer time and in communication with the peripherals. In this case, erroneous behavior obtained from trace buffer data (chip) or design and expected behavior obtained from specification constitute a counterexample. The AD function as introduced in Eq. (4.5) localizes the bug on the design:

$$\mathscr{F} = AD(Spec, \ Design, \ CEs) \qquad (4.6)$$

In this case, diagnostic traces are validated by the specification.

4.3.2   Case 2: Electrical Fault

The extracted responses from the trace buffer are not reproduced in the design while the responses of the design and the specification are consistent (Fig. 4.3, case 2). We assume that the silicon implements the RTL correctly. Therefore, in this case, there is an electrical fault which can be reproduced neither in the design nor the specification. One or more counterexamples are constituted by the erroneous behavior obtained from trace buffer data (chip) and the expected behavior produced by the design or the specification. Then, the AD function operates on the design (as RTL or gate level) to localize the root cause of electrical faults:

$$\mathscr{F} = AD(Chip, \ Design, \ CEs) \qquad (4.7)$$

For electrical faults, fault candidates are found on the design as this allows to access the internal structure of the circuit. Also, there is a mapping between the design and IC components. Diagnostic traces generated by DTG are checked by being applied to the chip. For electrical faults, the difficulty is to generate diagnostic traces which can reactivate the bug. Diagnostic traces can be generated by considering the suspected electrical fault type (e.g. drive strength, coupling, antenna effects, ...) and the layout information. Diagnostic traces can be applied instead of randomly generated traces [CMB07a] to improve the debugging performance. A diagnostic trace which leads to an inconsistency between the chip and the design is a counterexample. There are different methods for applying the diagnostic traces to a chip. Diagnostic traces can be applied to a chip by hardware structures like scan-chains or wrappers allowing what-if analysis [ABD+06]. *Software-Based Self-Testing* (SBST) methods are effectively used to apply test patterns in microprocessor-based systems [PGSR10]. In [CMAB09], a set of instructions, called *Access-Control Extensions* (ACE), are defined and used to access and to control the microprocessor's internal state. Directed tests on the hardware are run by utilizing ACE instructions.

4.3.3 Case 3: Electrical Fault and Design Bug

This case is a rare case in practice. When the extracted traces create different behaviors in the design in comparison to the extracted behavior of the chip, an electrical fault has occurred. In the case (Fig. 4.3, case 3), also the extracted traces create different behavior in the design and the specification which shows a design bug. Here debugging the design bug and the electrical fault can be performed independently and in parallel. Calling the functions defined by Eqs. (4.5) and (4.7) can discover the root causes of the design bug and the electrical fault concurrently:

$$
\begin{cases}
\mathscr{F}_D = AD(Spec, Design, CEs_D) \\
\mathscr{F}_E = AD(Chip, Design, CEs_E)
\end{cases}
\tag{4.8}
$$

The set CEs_D is the set of counterexamples for design bugs. The set CEs_E is the set of counterexamples for electrical faults. Also, the set $\mathscr{F}_D$ is the set of *design fault candidates*, i.e., fault candidates pointing to potentially buggy statements, and the set $\mathscr{F}_E$ is the set of electrical fault candidates, i.e., fault candidates that point to potentially faulty circuitry with behavior deviating from the gate-level specification.

4.3.4 Case 4: Indefinite Case

After the error detection on a chip, the extracted traces from the trace buffer create no inconsistency in any level. In this case (Fig. 4.3, case 4), the erroneous behavior related to the bug or fault activation time may not be stored in the trace buffer and is overwritten. To overcome this problem some approaches have been proposed in the cases of electrical faults and design bugs. In [dPNN+11], to get and to backspace an execution trace of on-chip signals for many cycles leading up to the activation time of a design bug, the chip is re-run with new trigger conditions. In each run, the state of the trace buffer is dumped out. This procedure repeats until the traces related to the activation time of the design bug can be extracted. Quick error detection and localization mechanisms are used to efficiently debug the electrical faults in [THPM+10] and [PHM09].

Another situation which may result in case 4 is a software bug or a bug related to hardware/software integration. The bugs in this case may be distinguished by running N-version programs (like software fault tolerant systems) and analyzing their behavior on software and hardware assertions.

4.4 Instantiation of the Generalized Debug Flow for Design Bugs

In this section, we show the application of the generalized function AD using case 1 when a design bug occurs in the system. Exemplarily we present AD for post-silicon debugging. Pre-silicon debug of Chap. 3 is also an instance of the generalized function AD. However, for post-silicon debugging, we need an additional step called trace analysis. By considering only design bugs, the automated debugging flow of Fig. 4.4 is instantiated from the generalized function AD. This flow includes the functions MBD (debugging), DTG, and DTV. Additionally, the trace analysis step is used as an initial step to obtain an initial counterexample. In the trace analysis step, trace buffer data which is obtained after running a test program on the chip should be analyzed and compared with the expected correct outputs obtained from the specification. As we consider only design bugs, the design and the chip have same behavior. If there is an inconsistency between trace data and golden data, this inconsistency or erroneous behavior represents a counterexample. By having the initial counterexample, the generalized AD (Eq. (4.5)) is invoked to execute MBD, DTG, and DTV (Eqs. (4.2)–(4.4)). Here we use SAT-based debugging as an effective approach to MBD. DTV is performed by applying the diagnostic traces to a simulation model or a formal model of the specification. DTG uses the MSPI method explained in Sect. 3.4.

4.5 Experimental Results

The effect of automated post-silicon debugging of design bugs on diagnosis accuracy, time, and memory is presented in this section. The hardware structure is written at RTL with Verilog hardware description language. The experiments are run on the

Fig. 4.4 Automated
post-silicon debugging
of design bugs

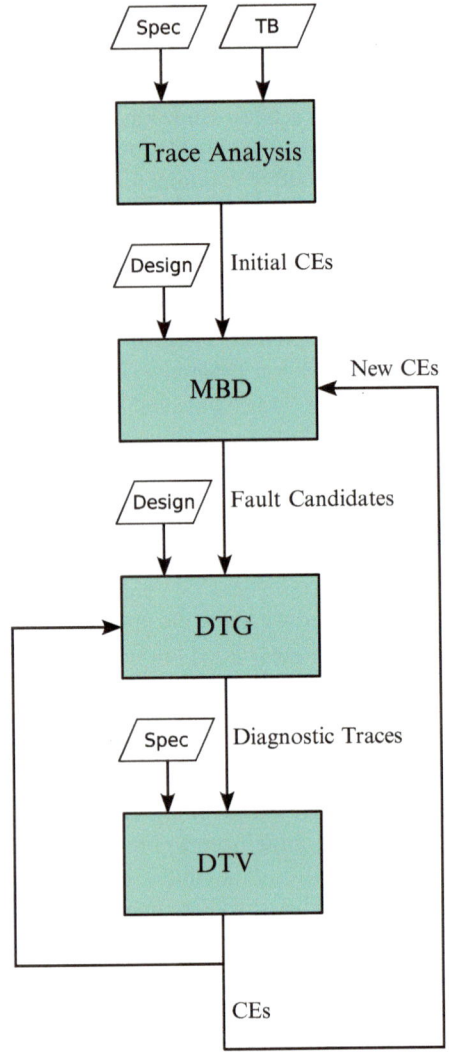

modules of the OpenRISC CPU from OpenCores [LB14]. A matrix multiplication
program is utilized as a test program to be run by OpenRISC in the ModelSim
environment. The experiments are executed for each module independently. We
insert a random single logic bug into the RTL code for each experiment. The logic
bugs constitute the largest fraction of the design bugs [CMA08] where the design
bugs are classified into logic bugs, algorithmic bugs and synchronization bugs.

The trace data related to a time window is recorded in the trace buffer of the
corresponding module. The time window is set to be 8 cycles. Each window contains
initial states at the first step of the window, inputs, and output results at the end of
the window. The size of the trace buffer is different for different modules, but the
maximum size is assumed to be $8\,K \times 32$ bits.

For specification, we use the Verilog modules as a black box module giving access only to module inputs, module outputs, and some of the internal registers (states) which would be available in a high level specification, too. For each time window, the recorded initial states and inputs are applied to the RTL design. Then, output results are compared to the output results of the corresponding window in the trace buffer to detect inconsistencies and to constitute initial counterexamples.

After having initial counterexamples, the RTL design is unrolled for 8 time steps for debugging. The techniques described in the chapter are implemented using C++ in the WoLFram environment [SKF$^+$09]. The visualization is performed by RTLvision PRO [Con14].

In the experiments, we compare the heuristic method for diagnostic trace generation (MSPI method) to a method based on random trace generation. In these experiments the methods are limited to a maximum of five iterations between the debugging and the verification procedures. The number of generated traces is limited to 2000 traces. An experiment terminates when either the maximum of five iterations reaches or the number of generated traces reaches the limit 2000.

The experimental results with respect to the diagnosis accuracy are presented in Table 4.1. The first and second columns show the module name of OpenRISC and the total number of components ($\#C$) which are used for SAT-based debugging. The third and fourth columns present the debugging result in the first session when debugging tries to find the potential number of fault candidates ($\#FC$) with the initial counterexamples ($\#CE$). The diagnosis accuracy is considered to be the inverse of $\#FC$. The result when the heuristic method is used for generating the diagnostic traces is shown in the columns 5–7, while the columns 8–10 are related to the random trace generation. The bold numbers highlight the best results in Table 4.1.

For $or1200\_alu$, the heuristic method obtains four new counterexamples after generating 41 diagnostic traces. Thus by a total number of five counterexamples (one initial counterexample and four new counterexamples) the diagnosis accuracy increases, i.e., the total number of fault candidates decreases compared to the initial set. Also the random method has the same accuracy for $or1200\_alu$. For $or1200\_ctrl$, $or1200\_genpc$, $or1200\_if$, $or1200\_operandmuxes$,

Table 4.1 Diagnosis accuracy

Method		Initial Result		Heuristic Method			Random Method		
Circuit	#C	#FC	#CE	#FC	#CE	#Trace	#FC	#CE	#Trace
or1200_alu	436	5	1	3	5	41	3	7	153
or1200_ctrl	1865	7	6	7	13	12	7	6	2000
or1200_genpc	732	20	3	**12**	8	296	20	3	2000
or1200_if	463	5	3	**1**	9	6	5	3	2000
or1200_lsu	793	3	2	**2**	9	10	3	7	2000
or1200_operandmuxes	376	11	3	**7**	8	62	11	3	2000
or1200_wbmux	278	7	10	7	15	13	7	10	2000

and *or1200_wbmux*, the random method cannot obtain any counterexample, while the heuristic method obtains the new counterexamples with a small number of diagnostic traces. For *or1200_lsu*, the random method generates some counterexamples which do not have potential to reduce the number of fault candidates, while the counterexamples obtained from diagnostic traces reduce the number of fault candidates. Totally, in four experiments out of seven experiments, the heuristic method achieves a better diagnosis accuracy than the random method.

The required run time (*Time*), and the maximum memory consumption (*Memory*) for each method is shown in Table 4.2. The time is related to debugging time and verification time. The bold numbers mark the best total times. The heuristic method spends less verification time than the random method in most of the experiments. In most cases some new counterexamples are identified after generating a small number of diagnostic traces.

The initial fault candidates for *or1200_if* are shown in Fig. 4.5. First, debugging finds five fault candidates corresponding to the three initial counterexamples. The fault candidates are highlighted with red color. Also the sections of code related to fault candidates are shown. After automation and by creating the new counterexamples, the number of fault candidates decreases. The schematic view of the circuit and its related code after automated debugging are shown in Fig. 4.6. Therefore, the designer focuses on a small set of fault candidates and can rectify the bug easily.

Overall, the experimental results show that debugging automation by using diagnostic traces increases the diagnosis accuracy and decreases the debug time.

4.6   Summary

This chapter presented an approach for automating debugging which can be used in different debugging scenarios from pre-silicon to post-silicon. The approach integrates model-based diagnosis, diagnostic trace generation, and diagnostic trace validation.

The proposed debugging flow can be applied to electrical faults as well as design bugs which slip into the IC from different levels of the system description. An instantiation of the generalized automated debugging flow was applied to post-silicon debugging of design bugs. The experimental results on post-silicon debugging showed that automated debugging by using diagnostic traces increases diagnosis accuracy and decreases debug time.

The fault candidates are visualized on the schematic view of the circuit. The fault candidates together highlight the sensitized paths leading to the erroneous behavior. Each counterexample helps the designer to understand the root cause of a bug by highlighting the new sensitized paths. Using more counterexamples decreases the number of fault candidates. In this case, the designer can concentrate only on a small set of fault candidates to fix the bug.

Table 4.2 Time and memory

Method		Heuristic Method				Random Method			
Parameter		Time			Memory	Time			Memory
Circuit	#C	Deb. (s)	Ver. (s)	Total (s)	(MB)	Deb. (s)	Ver. (s)	Total (s)	(MB)
or1200_alu	436	8.1	16.51	24.61	54	11.1	**6.56**	17.66	81
or1200 ctrl	1865	2950.24	**84.93**	3035.17	2332	356.99	240.95	597.94	1155
or1200_genpc	732	481.66	399.23	880.89	1154	101.55	**121.46**	223.01	337
or1200_if	463	116.43	**7.18**	123.61	652	26.2	76.93	103.13	166
or1200_lsu	793	275.61	**19.12**	294.73	1163	166.73	152.08	318.81	876
or1200_ oper-anrimiixes	376	124.29	**40.89**	165.18	330	22.52	76.52	99.04	144
or1200_wbmux	278	87.27	**6.96**	94.23	865	22.49	59.48	81.97	576

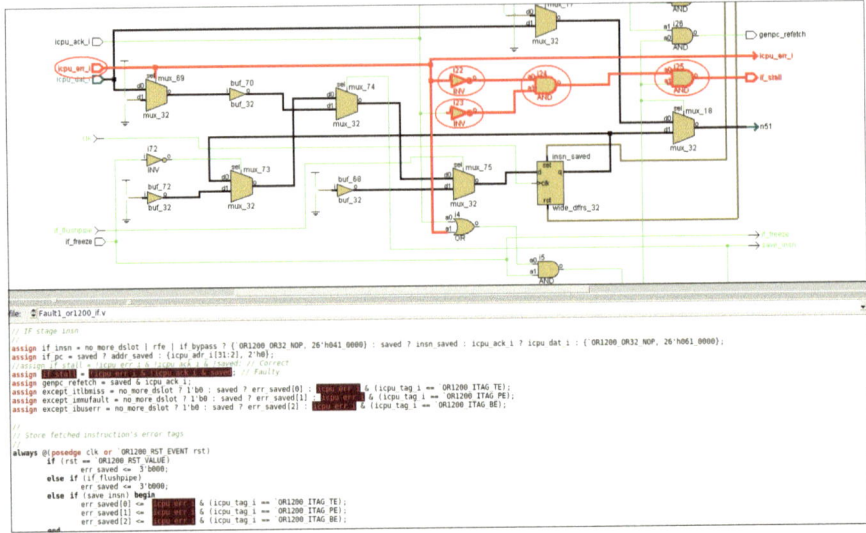

Fig. 4.5 Initial fault candidates

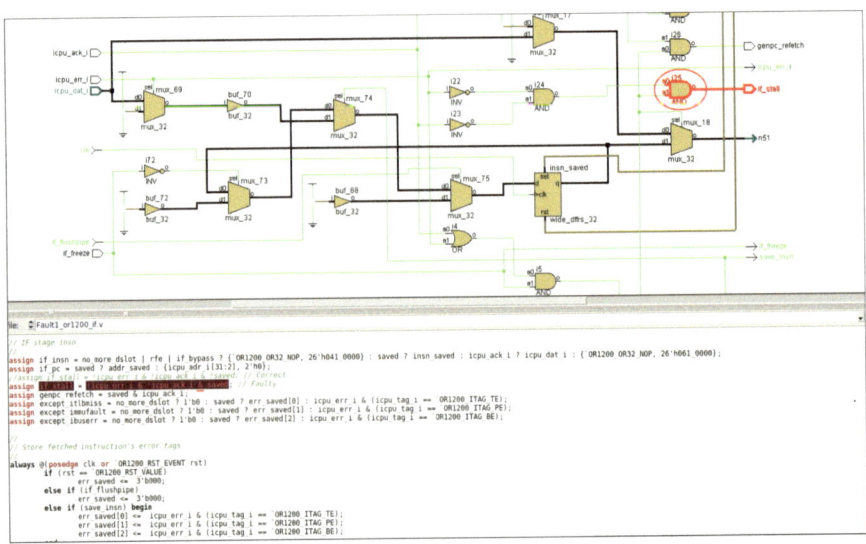

Fig. 4.6 Fault candidates after automation

In the next chapter, we develop an automated approach to debug synchronization bugs at pre-silicon stage.

Chapter 5
Automated Debugging for Synchronization Bugs

Design bugs at RTL are classified into three major classes: *logic bugs*, *algorithmic bugs* and *synchronization bugs* [CMA08]. There is a range of approaches to auto-mate the debugging process for logic bugs [SVAV05, CMB07b, SFD10, SFB$^+$09]. Algorithmic bugs often have a severe impact on the correctness of a design. Multiple major modifications are usually required to fix algorithmic bugs. Synchronization bugs are related to synchronization of data with respect to clock cycles in a design. For most of the synchronization bugs, a signal requires to be latched a cycle earlier or a cycle later in order to keep the correct timing behavior of signals in the design [CMA08]. To rectify the erroneous behavior caused by synchronization bugs, the designer debugging the erroneous behavior requires to manually add a flipflop or to manually remove a flipflop satisfying the correct timing behavior of the circuit. These bug models are called *missing flipflop* and *extra flipflop*.

Verification tools verify the behavior of a design against its specification in the pre-silicon stage. The correct timing behavior of a design is described by a spec-ification. The work in [CKY03] presents a formal method to specify the relations between multiple clocks and to model the possible behaviors. Then, a hardware design is verified against the specified clock constraints. An efficient clock modeling approach is presented in [GG07] to handle clock related challenges uniformly. The approach converts multiple clocks with arbitrary frequencies and ratios, gated clocks, multiple phases, latches, and flipflops in multi-clock synchronous system, into a single-clock model. Clock constraints are automatically generated to avoid unnecessary unrolling and loop-checks in BMC.

As mentioned earlier a model based on Boolean satisfiability is proposed in [SVAV05] to automate debugging of logic bugs. A circuit is enhanced with correction block in order to find the potential fault candidates. Abstraction and refinement techniques are used in [SV07] for handling the automated debugging of large designs with a better performance and reduced memory consumption. The work in [CMB07b] uses randomly generated counterexamples for debugging and applies automatic correction based on re-synthesis. An exact debugging approach

© Springer International Publishing Switzerland 2015
M. Dehbashi, G. Fey, *Debug Automation from Pre-Silicon to Post-Silicon*,
DOI 10.1007/978–3-319-09309-3_5

based on *Quantified Boolean Formulas* (QBF) is proposed in [SFD10] that creates high quality counterexamples to find fault candidates fixing any erroneous behavior. We introduced a pre-silicon debugging flow for testbench-based verification environments in Chap. 3. The approach uses diagnostic traces to obtain more effective counterexamples and to increase the diagnosis accuracy. All of the mentioned works consider logic bugs in order to automatically localize and to rectify an erroneous behavior at the pre-silicon stage.

In this chapter, we present an approach to automate the debugging of synchronization bugs at RTL [DF14a]. This is a class of bugs occurring while coding RTL. The approach of this chapter is an enhancement for the function MBD presented in Chap. 4 to cover synchronization bugs. First, synchronization bugs (extra/missing flipflop) are modeled and converted into a Boolean satisfiability formula. Having a counterexample given by a verification tool, our approach automatically extracts potential fault candidates which explain the erroneous behavior of the corresponding counterexample. Insertion or removal of some flipflops in the circuit can fix the erroneous behavior. By this, the approach automatically investigates the number of cycles that a signal needs to be latched earlier or later fixing the erroneous behavior of the circuit. In particular, our approach distinguishes synchronization bugs and logic/algorithmic bugs. The effectiveness and the diagnosis accuracy of our approach are shown in the experimental results.

The remainder of this chapter is organized as follows. The bug model is explained in Sect. 5.1. Our debugging approach is presented in Sect. 5.2. Section 5.3 explains how to model the correction of synchronization bugs in order to automatically debug a design using Boolean satisfiability. Then, the debugging algorithm is demonstrated in Sect. 5.4. Section 5.5 presents experimental results on benchmark circuits. Section 5.6 summarizes this chapter.

5.1 Synchronization Bug Model

Synchronization bugs occur when a signal is latched a cycle earlier or a cycle later in comparison to correct timing constraints. When a signal needs to be latched a cycle later, this behavior indicates lacking a flipflop on the corresponding signal. This bug is called *missing flipflop*. When a signal requires to be latched a cycle earlier, this may be due to an additional flipflop on the corresponding signal. This bug is called *extra flipflop*.

Example 5.1. Figure 5.1 shows synchronization bugs in a hardware system. The correct system is shown in Fig. 5.1a. The correct system has three main blocks A, B, and C. In each block, some flipflops are used to synchronize not only the internal timing of the corresponding block but also the external timing of the corresponding block communicating with other blocks. In Fig. 5.1b, there is an additional flipflop. The dotted circle shows the place of the bug in the system. The missing flipflop bug is shown in Fig. 5.1c.

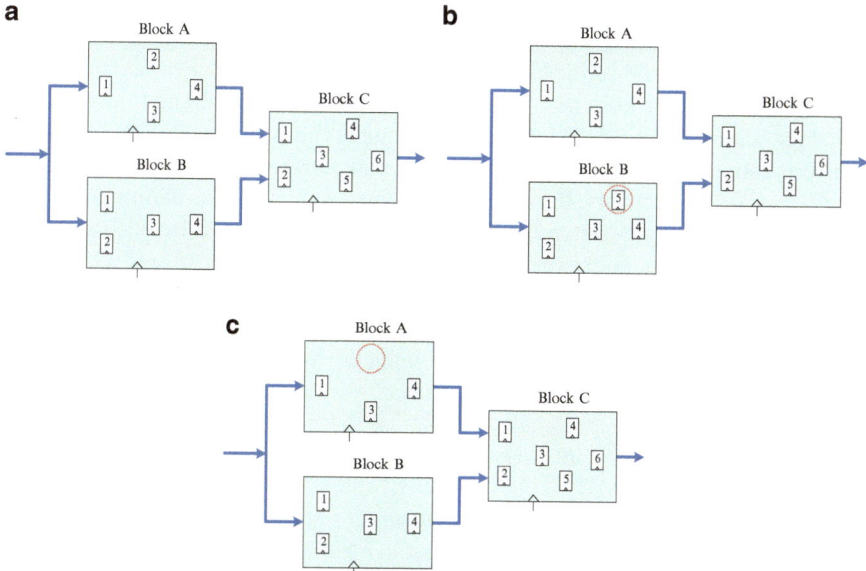

Fig. 5.1 Synchronization bug. (**a**) Correct system. (**b**) Extra flipflop bug. (**c**) Missing flipflop bug

In the case of a missing flipflop on signal s, the correction of this bug is adding one flipflop on the signal in order to postpone the propagation of the corresponding signal one clock cycle. The correction of this synchronization bug is shown as $s_2 =^{+1} s_1$, where $=^{+1}$ is the correction flipflop, and signal s is decomposed to signals s_1 and s_2 in order to add a flipflop on the propagation path of the signal. If signal s requires to be latched m clock cycles later, the correction is denoted as $s_2 =^{+m} s_1$, which indicates adding m consecutive flipflops on the propagation path of signal s.

In the case of an extra flipflop $s_2 = FF(s_1)$ in the circuit, deleting the corresponding flipflop causes the signal to be latched a clock cycle earlier. In this case, the correction is indicated as $s_2 =^{-1} s_1$, where $=^{-1}$ indicates a correction by removing the flipflop. By $s_2 =^{-m} s_1$ we denote removing m consecutive flipflops in order to latch a signal m clock cycles earlier.

5.2 Approach

Verification tools check the correctness of an implemented circuit according to the specification at the design step. If there is a contradiction between the behavior of the implemented circuit and the specification, this contradiction is returned as a counterexample.

Figure 5.2 shows our debugging approach. The approach is considered as an enhancement for the function MBD in Chap. 4. In this case, the initial counterexample is taken from the verification process. The function MBD is enhanced to have a two-stage debug process (two DBG functions in Fig. 5.2) helping to distinguish synchronization bugs and logic/algorithmic bugs. The first step in Fig. 5.2 shows the verification process. The outputs of the circuit and the specification are denoted by O_c and O_s, respectively. If the circuit and the specification have different output values, while the same test vector I is applied on inputs, this difference indicates a counterexample. A counterexample is shown by $CE(I, O)$, where $O = O_s$. A specification can be a formal specification, a simulation-based specification or golden data. We assume that all inputs and initial states are constrained by parameter I. Therefore, there is no free input or free initial state in the circuit. The verification function is written as follows:

$$CE = Verification\ (Circ,\ Spec) \tag{5.1}$$

Having a counterexample $CE(I, O)$, debugging starts. Here debugging uses the circuit and golden output values O_s in order to localize a bug.

First we create a new debugging instance without considering any bug model. The approach in [SVAV05] is utilized in order to localize a bug. In the approach, a multiplexer is inserted at the output of each component. When a multiplexer is activated, a new value is inserted at the output of the corresponding component fixing the erroneous behavior. This process is denoted as follows:

$$\mathscr{F} = DBG\ (ModelFree,\ Circ,\ CE) \tag{5.2}$$

A circuit, a counterexample CE and a parameter $ModelFree$ are the inputs of DBG. The parameter $ModelFree$ indicates that bug model-free debugging should be performed to extract the set of fault candidates $\mathscr{F}$. Bug model-free debugging was explained in Sect. 2.4.1.

Given the set of potential fault candidates $\mathscr{F}$, we investigate whether a fault candidate $FC \in \mathscr{F}$ can be a *synchronization fault candidate* according to the synchronization bug model. A *synchronization correction block* is inserted at the output of each fault candidate $FC \in \mathscr{F}$. The synchronization correction block changes the behavior of the fault candidate according to the synchronization bug model. Synchronization correction blocks are explained in Sect. 5.3 in detail. If by activating a synchronization correction block, the erroneous behavior of the circuit is fixed, a *synchronization fault candidate* FC' is detected. This process is denoted as follows:

$$\mathscr{F}' = DBG\ (Synch,\ Circ,\ CE,\ \mathscr{F}) \tag{5.3}$$

A circuit, a counterexample CE, a set of fault candidates $\mathscr{F}$ and a parameter $Synch$ are the inputs of DBG in this process. The parameter $Synch$ indicates that the

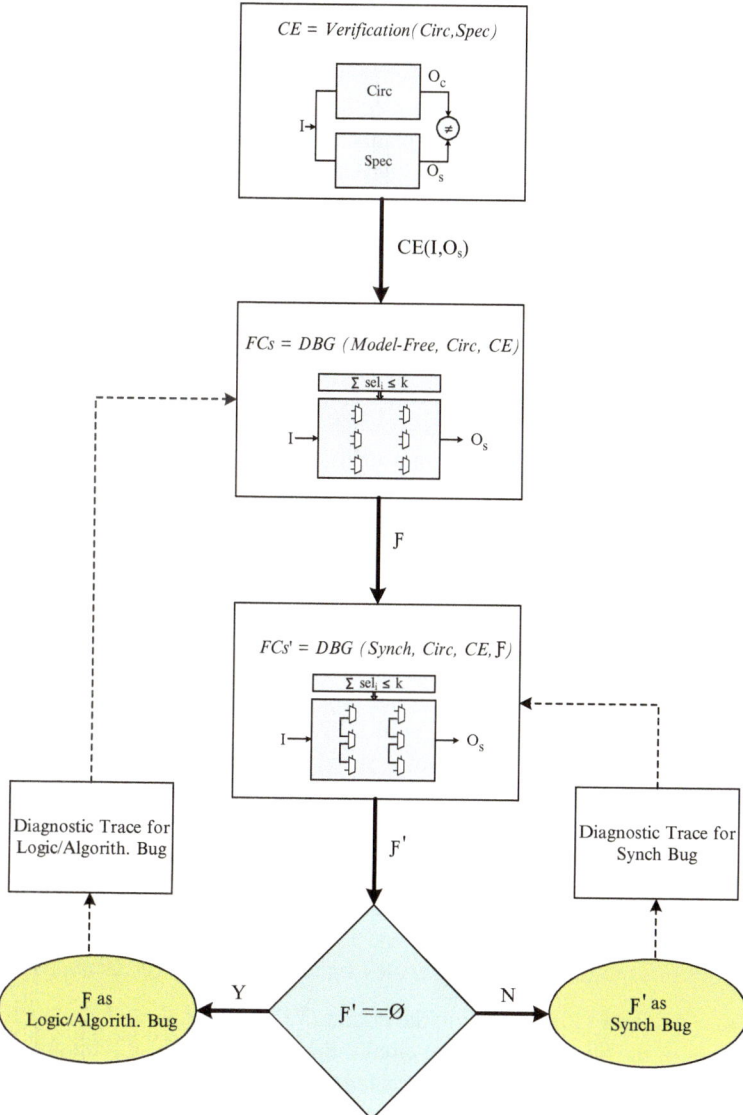

Fig. 5.2 Debugging approach

synchronization bug model is used for debugging. The output of the process is the set of synchronization fault candidates $\mathscr{F}' \subseteq \mathscr{F}$ where $FC' \in \mathscr{F}'$.

In the following steps of the debug flow, if the set of synchronization fault candidates is not empty, this set ($\mathscr{F}'$) is returned as fault candidates for synchronization bugs. Otherwise, the set of fault candidates $\mathscr{F}$ is returned as fault candidates for other kinds of bugs, i.e., logic bugs and algorithmic bugs.

Dashed lines in Fig. 5.2 indicate some additional processes in order to increase the diagnosis accuracy of debugging. Having the set of logic/algorithmic fault candidates, diagnostic traces can be generated (left dashed branch in Fig. 5.2). Diagnostic traces differentiate the behavior of fault candidates and help to create high quality counterexamples. High quality counterexamples help debugging to decrease the number of fault candidates. One approach to generate diagnostic traces was presented in Chap. 3. The approach does not need a bug model to generate diagnostic traces. Counterexamples obtained by diagnostic traces are used to iterate debugging and to increase the diagnosis accuracy.

In the case of synchronization fault candidates (right dashed branch in Fig. 5.2), *synchronization diagnostic traces* can be generated to distinguish the behavior of synchronization fault candidates. Synchronization diagnostic traces can be generated according to the synchronization bug model. Counterexamples obtained by synchronization diagnostic traces help debugging to decrease the number of synchronization fault candidates. Diagnostic traces can be generated similar to other fault models [LLC07]. The focus of this chapter is only on the synchronization bug model and debugging algorithm without using diagnostic traces.

5.3   Synchronization Correction Block

For debugging, first the circuit is unrolled as many times as the number of clock cycles constituting the corresponding counterexample. For example, if the length of the counterexample is three clock cycles, the circuit C is unrolled three times: C_0, C_1 and C_2. In this case, the input of a flipflop from clock cycle i is connected to the appropriate gates in clock cycle $i + 1$.

Example 5.2. In Fig. 5.3, there is a circuit including two flipflops A and B. To do the debugging, the circuit is unrolled three times. Flipflops A and B are removed. The input wire of flipflop B at cycles 0 and 1, b_0 and b_1, are connected to the output wire of the corresponding flipflop at cycles 1 and 2, c_1 and c_2, respectively.

There is a missing flipflop bug in the circuit of Fig. 5.4. A dotted circle indicates the location of the bug. To debug the circuit, three copies of the circuit are created. The input of flipflop A at cycle i is connected to the appropriate signal at cycle $i + 1$. For debugging, we investigate at which wire of the circuit a flipflop is missing. Therefore, we need a correction block at each wire of the circuit which is able to model the behavior of a flipflop at the corresponding wire. The green part in Fig. 5.4 shows this model. Multiplexers are utilized to model a flipflop behavior at every wire of the circuit. When there is a set of fault candidates $\mathscr{F}$ given by bug model-free SAT-based debugging, correction blocks are inserted only on each fault candidate $FC \in \mathscr{F}$. To model correction block for m consecutive missing flipflops, m consecutive correction blocks can be inserted on each wire.

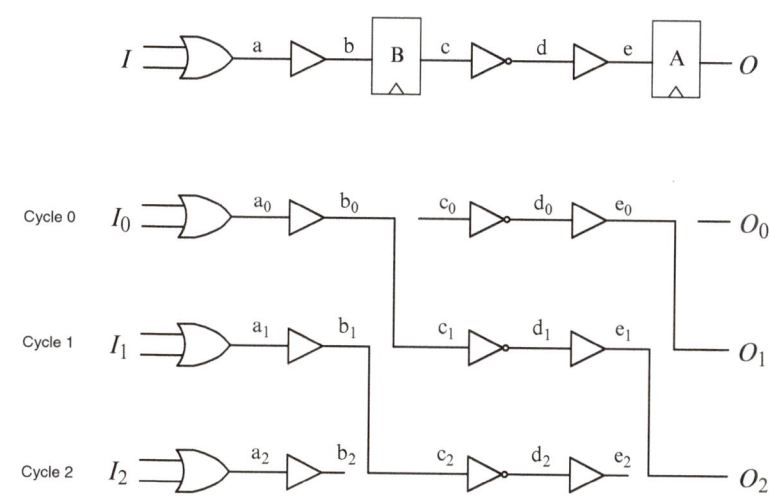

Fig. 5.3 Unrolled correct circuit

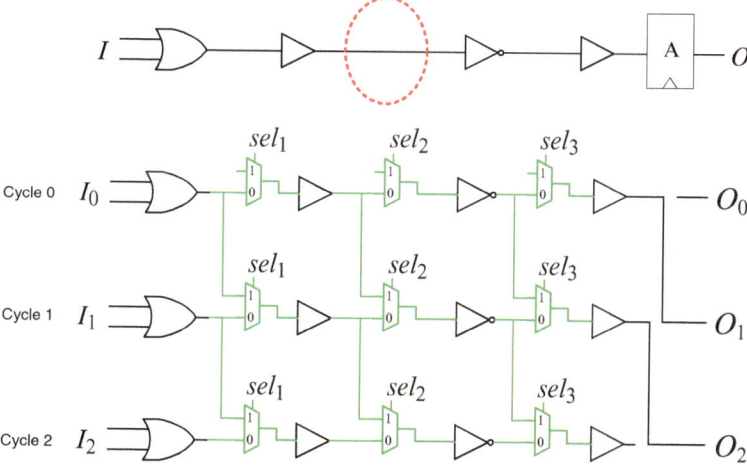

Fig. 5.4 Debugging instance for missing FF bug

A flipflop behavior at wire p is activated if select line sel at wire p is active ($sel = 1$). Therefore, the output of the activated flipflop at cycle i is connected to the input of the corresponding flipflop at cycle $i - 1$. If the correction block is inactive ($sel = 0$), the circuit at wire p has the behavior of a normal wire.

Having one correction block at every wire of the circuit, the inputs and output of the model are constrained according to the inputs and output values of the corresponding counterexample. Then debugging answers the following question by activating the select lines: If a flipflop is added at wire p, can the erroneous behavior of the corresponding counterexample be fixed? In this case, a SAT solver is utilized to extract all possible fault candidates. The number of correction blocks could be

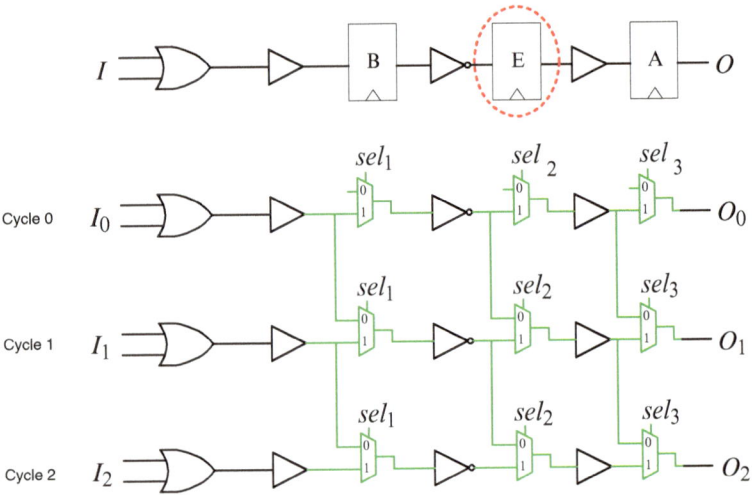

Fig. 5.5 Debugging instance for extra FF bug

optimized, e.g., by saving correction blocks around buffers and inverters where the insertion of a flipflop before or after the gate is functionally equivalent.

The model for the extra flipflop bug is shown in Fig. 5.5. A correction block is applied at the location of every flipflop in the circuit. The correction block for the extra flipflop bug has the reverse behavior in comparison to a correction block for the missing flipflop bug. For this kind of bugs, debugging investigates if removing a flipflop can fix the erroneous behavior of the corresponding counterexample. If there is a set of fault candidates $\mathscr{F}$ given by bug model-free SAT-based debugging, correction blocks are inserted only on flipflops which are fault candidates in set $\mathscr{F}$.

When there is both a missing flipflop bug and an extra flipflop bug in a design, correction blocks for both kinds of bugs are required simultaneously. In this case, a missing flipflop correction block is inserted on every wire of the circuit, while extra flipflop correction blocks are inserted only at existing flipflops. A constraint on the select lines of the correction blocks controls the whole synchronization variation.

We add the following constraint to the created debugging instance to control the select lines of multiplexers for single bugs:

$$\sum_{i=1}^{n} sel_i \leq 1 \qquad (5.4)$$

Parameter n is the total number of correction blocks in the debugging instance. For multiple bugs, k bugs may be activated at the same time. In this case, the following constraint controls the behavior of multiplexers (correction blocks):

$$\sum_{i=1}^{n} sel_i \leq k \qquad (5.5)$$

Fig. 5.6 Synchronization
debugging algorithm

```
1   algorithm DBG (In : Synch, Circ, CE, ℱ, Out : ℱ')
2   // ℱ' = ℱ'_Extra ∪ ℱ'_Missing ∪ ℱ'_Mixed
3   // Maximum number of synchronization bugs : k_max
4
5   {SEL_E, SEL_M} = Insert_Extra_Missing_CLs(Circ, ℱ)
6   SEL = SEL_E ∪ SEL_M
7   // sel_i ∈ SEL, i = 1, 2, ..., n
8   Add_Constraint(CE)
9
10  {ℱ'_Extra, ℱ'_Missing, ℱ'_Mixed} = {∅, ∅, ∅}
11  k = 1
12  do
13  {
14      Add_Constraint(∑_{i=1}^{n} sel_i ≤ k)
15      if Solve() == SAT then
16      {
17          Solutions = Extract_All_Solutions()
18          break
19      }
20      else
21      {
22          Remove_Constraint(∑_{i=1}^{n} sel_i ≤ k)
23          k = k + 1
24      }
25  } while k ≤ k_max
26
27  foreach Sol ∈ Solutions do
28  {
29      if ∀sel_i ∈ Sol : sel_i ∈ SEL_E then
30          ℱ'_Extra = ℱ'_Extra ∪ Sol
31      else if ∀sel_i ∈ Sol : sel_i ∈ SEL_M then
32          ℱ'_Missing = ℱ'_Missing ∪ Sol
33      else
34          ℱ'_Mixed = ℱ'_Mixed ∪ Sol
35  }
36
37  ℱ' = {ℱ'_Extra ∪ ℱ'_Missing ∪ ℱ'_Mixed}
38  end algorithm
```

5.4 Algorithm

The algorithm of our synchronization debugging is shown in Fig. 5.6 in pseudocode. The inputs of the debugging algorithm are a parameter $Synch$, a circuit, a counterexample CE and a set of fault candidates $\mathscr{F}$ given by bug model-free SAT-based debugging. The parameter $Synch$ indicates that the algorithm is a synchronization debugging algorithm. The output of the algorithm is a set of synchronization fault candidates $\mathscr{F}'$. The set $\mathscr{F}'$ is a union of three sets $\mathscr{F}'_{Extra}$, $\mathscr{F}'_{Missing}$ and $\mathscr{F}'_{Mixed}$ (line 2). Set $\mathscr{F}'_{Extra}$ represents the set of extra flipflop fault candidates. Set $\mathscr{F}'_{Missing}$ represents the set of missing flipflop fault candidates. When a fault candidate is composed of multiple components such that some components of the fault candidate are missing flipflops while other components of the fault candidate are extra flipflops, this fault candidate is included in set $\mathcal{F}'_{Mixed}$. In the algorithm, the maximum number of synchronization bugs is given by parameter k_{max} (line 3) which limits variable k in Eq. (5.5).

At the first step of the algorithm, correction blocks for extra flipflop and missing flipflop models are inserted at the locations of fault candidates $FC \in \mathcal{F}$ (line 5). The set of select lines for extra flipflop correction blocks is denoted by set SEL_E. Set SEL_M denotes the set of select lines for missing flipflop correction blocks. The whole set of select lines is stored in set SEL (lines 6–7). Line 8 constrains the inputs and the output of the created instance to the values given by the counterexample CE. In line 10, the sets $\mathcal{F}'_{Extra}$, $\mathcal{F}'_{Missing}$ and $\mathcal{F}'_{Mixed}$ are initialized. In line 11, variable k is initialized to 1. Variable k indicates the number of bugs in the design. Initially, the algorithm starts with the assumption of having a single bug ($k = 1$). The constraint of line 14 controls the select lines of multiplexers. Then a SAT solver is called to find a solution (line 15). If there is any solution, all solutions will be extracted (line 17). The solutions are saved in set $Solutions$. Otherwise, the previous constraint of select lines is removed (line 22) and k increases (line 23). If variable k is less than or equal to the parameter k_{max} (line 25), the algorithm iterates until finding a solution or reaching k_{max}.

After finding solutions, the solutions are categorized into three sets $\mathcal{F}'_{Extra}$, $\mathcal{F}'_{Missing}$ and $\mathcal{F}'_{Mixed}$ (lines 27–35). If all components of a solution Sol are included in set SEL_E (line 29), the fault candidate is considered as extra flipflop fault candidate (line 30). If all components of a solution Sol are included in set SEL_M (line 31), the fault candidate is considered as missing flipflop fault candidate (line 32). Otherwise, the fault candidate has some extra flipflop components and some missing flipflop components and is added to set $\mathcal{F}'_{Mixed}$ (lines 33–34).

5.5 Experimental Results

The effects of our debugging approach are experimentally demonstrated in this section on sequential circuits of LGSynth-93 and ITC-99 benchmark suites and RTL modules of the OpenRISC CPU. We inject the single synchronization bugs randomly by removing a flipflop (Missing FF) or by adding a flipflop (Extra FF). The single logic bugs are randomly injected by replacing gates. For example an AND gate is replaced by an OR gate. For bounded sequential debugging, the circuits are unrolled up to thirty time steps.

The experiments are carried out on a Dual-Core AMD Opteron(tm) Processor 2220 SE (2.8 GHz, 32 GB main memory) running Linux. We use MiniSAT as underlying SAT solver [ES04]. The techniques described in this chapter are implemented using C++ in the WoLFram environment [SKF+09]. Run time is measured in CPU seconds, and the memory consumption is measured in MB.

Having a buggy circuit, the debugging approach of Fig. 5.2 is called. The verification process returns an initial counterexample. Then, bug model-free SAT-based

Table 5.1 Experimental results (diagnosis accuracy)

Benchmarks			Missing FF		Extra FF		Logic Bug	
Circuit	#Gates	#FF	#FC	#FC'	#FC	#FC'	#FC	#FC'
b01	49	5	8	3	8	1	7	0
b02	25	4	13	3	6	1	16	0
b04	707	66	26	9	36	12	18	0
b05	1054	34	22	7	10	2	2	0
b08	177	21	39	11	13	2	6	0
b10	211	17	55	5	4	1	3	0
b11	790	31	43	10	49	9	8	0
b12	1062	121	68	11	61	7	114	0
gcd	1012	59	10	6	3	1	6	0
phase_de.	1672	55	20	10	115	24	29	0
or1200_ctrl	5093	198	7	3	6	1	7	0
or1200_if	2029	69	7	4	8	3	6	0
or1200_lsu	1777	8	6	3	21	5	10	0
or1200_operand.	1485	66	13	6	14	7	12	0

debugging searches for the potential fault candidates $\mathcal{F}$. Set $\mathcal{F}$ is given to the synchronization debugging in order find the synchronization fault candidates $\mathcal{F}'$. If set $\mathcal{F}'$ is empty, there are no single synchronization bugs according to the synchronization bug model.

The experimental results are presented in Table 5.1 for single faults. There are four sections in Table 5.1. Section *Benchmarks* shows the characteristics of the benchmarks (columns 1–3). Other sections show the experimental results when a missing flipflop, an extra flipflop or a logic bug is injected. The final number of fault candidates ($\#FC$) and the final number of synchronization fault candidates ($\#FC'$) are shown in each section. Table 5.2 shows the required run time (*Time*) and the maximum memory consumption (*Mem*) for our experiments. The time includes the verification time (*Ver.*), the time for bug model-free debugging (*DbgM*), the time for synchronization debugging (*DbgS*) and the total time (*Total*).

Consider the greatest common divider *gcd* in the table; when there is a missing flipflop in circuit *gcd*, bug model-free debugging finds ten fault candidates ($\#FC = 10$). These ten fault candidates are given to synchronization debugging. Then synchronization debugging investigates whether a fault candidate $FC \in \mathcal{F}$ can be a synchronization fault candidate according to the synchronization fault model. Synchronization debugging returns a set of synchronization fault candidates $\mathcal{F}' \subseteq \mathcal{F}$. For circuit *gcd*, six synchronization fault candidates ($\#FC' = 6$) are found by synchronization debugging. The number of fault candidates ($\#FC$) is by definition always larger equal than the number of synchronization fault candidates ($\#FC'$). This is one reason that causes the debugging time to extract set $\mathcal{F}$ (column *DbgM*) to be longer than the time to extract set $\mathcal{F}'$ (column *DbgS*).

Table 5.2 Experimental results (time and memory)

Benchmarks	Missing FF					Extra FF					Logic Bug				
	Time				Mem	Time				Mem	Time				Mem
Circuit	Ver.	DbgM	DbgS	Total		Ver.	DbgM	DbgS	Total		Ver.	DbgM	DbgS	Total	
b01	0.0	0.2	0.0	0.2	15.5	0.1	0.2	0.0	0.3	15.6	0.0	0.1	0.0	0.1	15.4
b02	0.0	0.2	0.1	0.3	15.6	0.0	0.1	0.0	0.2	15.5	0.0	0.2	0.0	0.3	15.6
b04	2.4	11.1	1.0	14.4	17.6	2.5	20.2	1.2	23.8	17.6	14.7	35.7	2.5	52.8	19.3
b05	0.7	6.7	0.5	7.9	16.5	0.7	4.3	0.3	5.3	16.4	0.7	2.3	0.2	3.2	16.3
b08	1.2	7.9	1.3	10.5	18.0	0.4	1.6	0.2	2.2	16.3	0.1	0.4	0.1	0.5	15.9
b10	0.5	11.1	0.6	12.3	17.3	0.1	0.4	0.1	0.6	15.8	0.1	0.6	0.1	0.7	15.8
b11	21.3	70.0	5.0	96.4	19.6	3.0	59.4	1.3	63.8	17.5	2.9	7.3	0.5	10.6	17.0
b12	7.4	92.1	2.1	101.6	19.3	8.8	82.8	1.7	93.3	19.2	42.3	488.9	8.7	539.9	24.1
gcd	0.7	6.4	0.5	7.5	16.4	8.1	16.5	0.9	25.6	17.9	17.0	36.0	2.8	55.7	19.1
phase_de.	18.5	41.8	4.5	64.7	19.3	61.6	904.3	17.7	983.6	21.9	15.5	86.5	2.4	104.4	18.9
or1200_ctrl	50.2	222.9	14.1	287.2	39.3	70.7	191.8	6.2	268.7	32.0	16.6	107.2	2.1	125.9	26.0
or1200_if	2.9	19.5	1.0	23.4	20.6	3.3	24.5	1.1	28.8	22.7	5.3	27.4	0.7	33.4	22.7
or1200_lsu	25.2	76.1	6.1	107.4	26.6	39.5	170.4	12.5	222.4	30.2	22.5	83.3	5.5	111.3	26.2
or1200_operand.	0.9	13.4	0.8	15.0	20.1	1.0	12.8	0.9	14.7	19.9	0.6	9.4	0.3	10.2	18.6

In most of the cases, there are fewer synchronization fault candidates for extra flipflop bugs in comparison to the missing flipflop bugs (comparison of column 5 and column 7 in Table 5.1). The reason is that flipflops may only be removed at places where they are located in the design. But adding of flipflops is possible at any signal. So the number of flipflops that can potentially be added is much larger than the number of flipflops that can potentially be removed.

When there is a logic bug in circuit $b01$, the number of fault candidates is seven ($\#FC = 7$), while the number of synchronization fault candidates is zero ($\#FC' = 0$). When there is no synchronization fault candidate, our algorithm can distinguish a logic bug from a synchronization bug. We detect that there is no synchronization bug in the circuit according to the synchronization bug model. In this case, the fault candidates $\mathscr{F}$ are returned as logic fault candidates as shown in Fig. 5.2.

In principle, a logic bug may be correctable by a synchronization correction block with respect to the given counterexample. But in the examples of Table 5.1, this case does not occur because of the length of counterexamples. The length of the counterexample, i.e. the number of clock cycles relevant to the counterexample, affects the accuracy of distinguishing between a logic bug and a synchronization bug. We show the relation of counterexample length and distinguishing logic bugs versus synchronization bugs for circuit $b01$ in Fig. 5.7. When there is a logic bug in circuit $b01$, some random counterexamples with different lengths are generated. The diagram shows the number of synchronization fault candidates is zero when a counterexample is longer (CE length $= 7$). Therefore in this case, we can distinguish a logic bug from a synchronization bug. Figure 5.8 shows the effect of counterexample length on distinguishing logic bugs versus synchronization bugs for circuit $b10$. In this case when the length of the counterexample is 10, we can distinguish a logic bug from a synchronization bug. Not only the length of a counterexample but also the number and the quality of counterexamples can influence the accuracy of the classification.

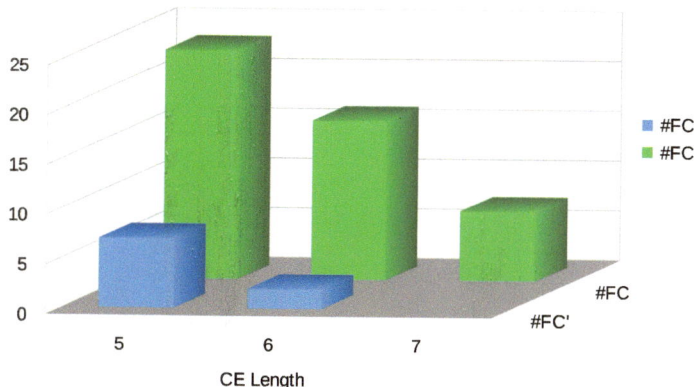

Fig. 5.7 Effect of CE length on distinguishing logic bug and synchronization bug (circuit b01)

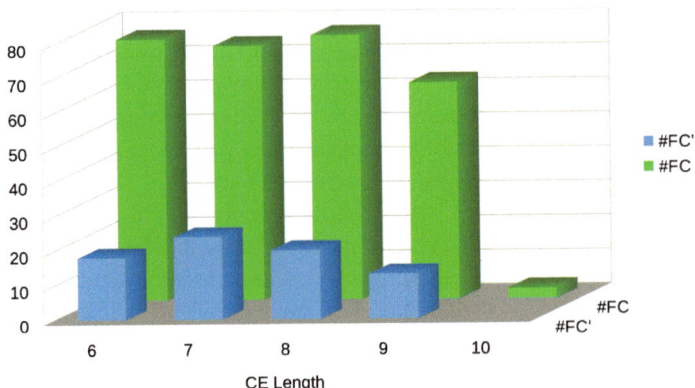

Fig. 5.8 Effect of CE length on distinguishing logic bug and synchronization bug (circuit b10)

5.6 Summary

In this chapter, we introduced an approach to automate debugging for synchronization bugs at RTL. The correction for synchronization bugs is modeled in order to automatically debug a design using SAT. The approach automatically investigates the number of cycles that a signal needs to be latched earlier or later fixing the erroneous behavior of the circuit.

Also we presented an approach to distinguish synchronization bugs and logic/algorithmic bugs. Our approach integrates bug model-free debugging and synchronization debugging in order to differentiate synchronization bugs from other bugs. The experimental results showed diagnosis accuracy and efficiency of the approach in distinguishing synchronization bugs and logic/algorithmic bugs.

The focus of this part was debugging of design bugs, i.e., logic bugs and synchronization bugs. The next part deals with the debugging for timing faults at post-silicon stage.

Part II
Debug of Delay Faults

Chapter 6
Analyzing Timing Variations

Variability is recognized to be a major challenge in analyzing the circuits as IC technology continues to scale down. In this case, delay deviations are imposed by process variations such as uncertainty in the parameters of fabricated devices and interconnects, and by environmental variations such as temperature and voltage [BCSS08, APP10, SGT+08].

Recently, there is a range of works that considers timing analysis of circuits under variations. A survey of the works focusing on *Statistical Static Timing Analysis* (SSTA) is given in [BCSS08]. The statistically-critical paths under process variations are extracted by a bound-based method in [XD11]. The extracted paths have the highest probability to fail the timing constraint. The works in [APP10, GYP+10, SGT+08] analyze the effects of process variations on the delays of logic gates and timing errors. Timing error detection and correction has been proposed as an approach to bridge the gap between typical case and worst case design, by allowing circuits to operate without any margins [EKD+03, TBW+09]. Variation-induced errors for power and processor frequency can be mitigated by microarchitectural techniques using the framework presented in [SGTT08].

On the other hand, in the recent years, significant progress in areas such as approximate and probabilistic computing has been achieved. Computations typically addressed in these areas focus on good enough or bounded results but not necessarily exact results [Bre10]. These techniques relax the requirement of exact numerical or Boolean equivalence to yield performance or energy efficiency [SAKJ10, CR10, VARR11].

An approximate implementation of a circuit does not exactly match the specification because of timing-induced errors or functional approximations [Bre10]. Timing-induced errors can be produced by voltage over-scaling or overclocking.

The works in [SG10, SG11, LEN+11] exploit systematic synthesis of approximate circuits to reduce circuit area and delay as well as to increase yield. A logic optimization procedure is developed in [WC09] that utilizes multi-V_t (threshold voltage) libraries to optimize a circuit for higher frequency and throughput under

© Springer International Publishing Switzerland 2015
M. Dehbashi, G. Fey, *Debug Automation from Pre-Silicon to Post-Silicon*,
DOI 10.1007/978-3-319-09309-3_6

timing error detection and correction. The work in [KKKS10] uses a power-aware slack redistribution that shifts the timing slack of frequently-exercised, near-critical timing paths in a power- and area-efficient manner. An *Error-Resilient System Architecture* (ERSA) is presented in [LCB$^+$10] which combines unreliable cores with a small fraction of reliable processor cores for running system software, controlling application flow, and recovering from errors generated on unreliable cores. Scalable effort hardware design is proposed in [CMR$^+$10] to identify mechanisms at each level of design abstraction (circuit, architecture, and algorithm) which can be used to vary the computational effort expended for generating the exact results. Utilizing these scaling mechanisms improves energy efficiency while maintaining an acceptable result.

The work in [VARR11] proposes a systematic methodology for the modeling and analysis of circuits for approximate computing. The methodology is utilized to analyze a circuit under timing-induced approximations as well as functional approximations using multiple metrics. However, timing variations are not considered. Since variations can significantly perturb the timing of various paths in a circuit, it is natural to expect that they will also significantly impact which paths fail under timing-induced approximations, and therefore, the functional behavior of approximate circuits. For this purpose considering the timing behavior of variations during the analysis of approximate circuits is essential.

A unified framework is proposed in this chapter [DFRR12] that can be used to analyze

- how a circuit behaves under timing variations,
- how a circuit behaves under timing-induced approximations, and
- how an approximate circuit behaves under timing variations.

By considering the functional domain, our approach is complementary to SSTA. In the framework, first the timing behavior of a circuit is converted into the functional domain according to a time unit model. The newly constructed circuit is called the *Time Accurate Model* (TAM) of the circuit. The TAM represents the functional behavior of the circuit with respect to the circuit delay and a precision of an arbitrarily fine-grained but discrete time unit. Afterwards, *Variation Logic* (VL) is inserted in the TAM to apply the timing variation. The VL is applied at each gate. The cumulative slowdown or speedup normalized to a time unit may affect the correct behavior of the circuit. A component called *Variation Control* (VC) determines the behavior of the VL. Moreover, the circuit is enhanced by *Time Control* (TC) logic. TC is a flexible model to control the frequency at the inputs. TC models timing-induced approximations like overclocking. For this work, SAT is utilized as an underlying reasoning engine to analyze the circuits.

The rest of this chapter is organized as follows. Section 6.1 introduces preliminary information on timing parameters. Section 6.2 describes our approach to construct the TAM, to insert VL, and to enhance a circuit by TC and VC. The properties of the TAM are discussed and proved in Sect. 6.3. First the properties of the TAM for the outputs at one time step are shown. Then we show how waveforms are modeled using the TAM. Experimental results on arithmetic units are presented in Sect. 6.4. Section 6.5 concludes this chapter.

6.1 Timing Parameters

The amount of time that a signal requires to propagate from component's inputs to its output is called *Delay* of the component. A change of the component delay is called *timing variation*. Timing variation may increase the component delay called *slowdown*, or may decrease the component delay called *speedup*.

A *time unit* is considered as an arbitrarily fine-grained but discrete unit of delay. The delays of gates and interconnects are assumed to be an integer multiple of one time unit.[1] In a circuit where the shortest path delay is D_s time units, and the longest path delay is D_l time units, the current output O_t depends on the inputs of $I_{t-D_s}, I_{t-D_s-1}, \ldots, I_{t-D_l}$ where indices denote the times of input with a step of one time unit. Each index is called *time step*.

T time units define a *clock period*. In synchronous circuits, the input to the combinational logic changes only once every *clock cycle*. The times of inputs with a step of one clock period are denoted by clock cycles. If the circuit has a clock period of T, the output at time step t depends on the inputs of the following clock cycles:

$$\forall i, a \leq i \leq b : [I_{t-iT-1}, \ldots, I_{t-(i+1)T}] \tag{6.1}$$

$$a = \lceil D_s/T \rceil - 1, \quad b = \lceil D_l/T \rceil - 1$$

The input times are partitioned by this formula according to the clock period T such that in each clock cycle, the inputs are assumed to be fixed. For example, when $D_s = 1, D_l = 5, T = 5, O_t$ depends on the input values from the time steps that fall within the previous clock cycle $[I_{t-1}, I_{t-2}, I_{t-3}, I_{t-4}, I_{t-5}]$, and in this clock cycle, the inputs do not change: $I_{t-1} = I_{t-2} = I_{t-3} = I_{t-4} = I_{t-5}$. When $T = 2, O_t$ depends on the input values from the time steps that fall within the following clock cycles: $[I_{t-1}, I_{t-2}], [I_{t-3}, I_{t-4}], [I_{t-5}, I_{t-6}]$. Overall, when $T < D_l$, the clock is overscaled, i.e., the current output depends on the inputs of multiple previous clock cycles. This case is called *overclocking*. We note that, for our purpose, voltage overscaling has the same effect as overclocking, since the delays of the gates will be scaled up based on lower voltage, while the clock period remains the same.

If the clock is overscaled, longer paths fail because the input does not have enough time to propagate to the output. In this case, the current output result depends not only on the input of one previous clock cycle but also on the inputs of multiple previous clock cycles. The "older" inputs (the inputs of the clock cycles more distant

---

[1]At the end of Sect. 6.2.3, we discuss how more complex timing models can be handled by our approach.

from current time t) influence the output through longer paths. The "newer" inputs (the inputs of the clock cycles closer to the current time t) affect the output through shorter paths.

6.2 Approach

We convert the fine-grained timing behavior of a circuit into the functional domain. Having the behavior of a circuit according to a fine-grained time unit allows us to utilize it for modeling races, glitches, etc. The fine-grained timing model is also utilized to control and to modify the frequency of a circuit during our analysis. This fine-grained timing model of the circuit is called TAM. When the timing behavior of a circuit is available in the functional domain, formal verification methods can comprehensively analyze the timing effects of the circuit.

The overview of our approach is shown in Fig. 6.1. First, the gate level circuit (synthesized netlist) is generated according to a cell library. Afterwards, the time accurate model of the circuit is created by the TAM engine. The TAM models the fine-grained timing behavior of a circuit in the functional domain. The TAM is generated according to a fine-grained time unit. The time unit specifies the granularity of the analysis. The delays of all gates in the circuit are normalized according to the time unit. VL is inserted in the TAM according to the maximum timing variation (D).

In the final step, TC and VC are added. TC consists of some constraints on the inputs to control the clock period according to Formula (6.1). VC is a constraint to

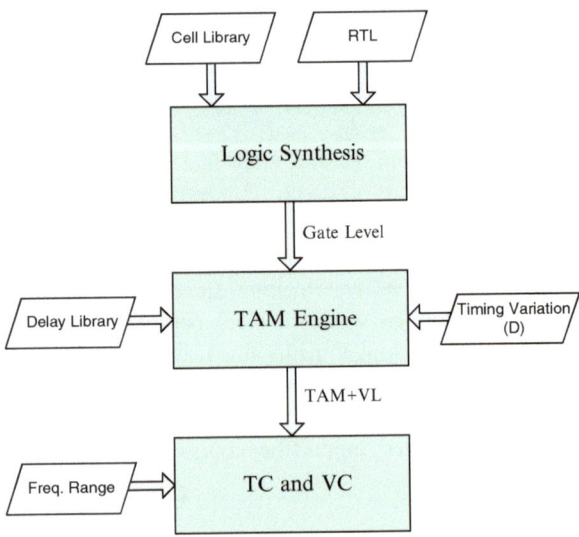

Fig. 6.1 Overview of the proposed approach

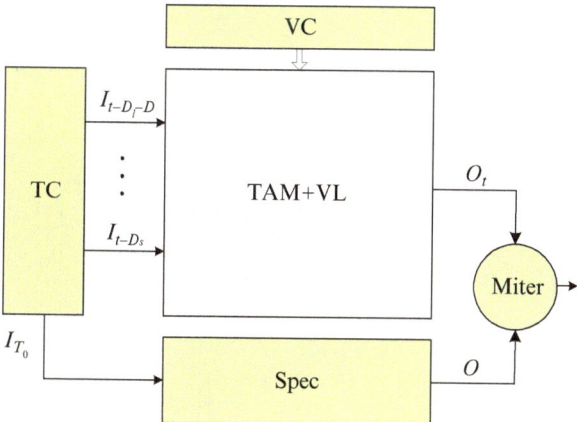

Fig. 6.2 Overall model created by the framework

control the timing variation. The final model can be used to analyze a circuit with respect to different frequencies.

The overall model created by our framework for the analysis is shown in Fig. 6.2. The main components of the model are: TAM+VL, TC, VC. Also the model includes two side components: *spec* and *miter*. These two components serve different tasks in different applications. *Spec* can be a golden specification or golden properties of the ideal circuit behavior. The deviation of the circuit output result against its specification is measured by *Miter*. Here, we use a SAT solver as an underlying engine to measure the deviations.

A designer can answer the following questions by using our framework while varying the frequency:

- How does a circuit behave under timing variations?
- How does a circuit behave under timing-induced approximations, i.e., under overclocking?
- How does an approximate circuit (overclocked circuit) behave under timing variations?

In the following, we describe the process to create the TAM and its VL in Sect. 6.2.1. The TC and VC components are explained in Sect. 6.2.2. Section 6.2.3 discusses how speedup variations are considered in our framework.

6.2.1 TAM Engine

The TAM algorithm was inspired from the algorithm presented in [VARR11]. However, there are key differences necessitated by the need to handle variations— we consider a fine-grained time unit and also the variation logic is added into the model.

Fig. 6.3 Converting original gates and wires to untimed gates and wires

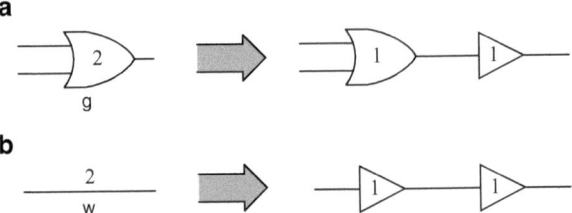

Fig. 6.4 Creation of time accurate model

```
1   algorithm  TAM (In : untimed circuit, Out : TAM circuit)
2   time = 0
3   SIG = PO
4   while  SIG ≠ ∅  do
5   {
6       SIG_temp = ∅
7       foreach  sig ∈ SIG  do
8       {
9           gate = predecessor(sig)
10          copy(gate, i_{t−time−1}, o_{t−time})
11
12          foreach  input ∈ I(gate)  do
13              if  input ∉ SIG_temp  and  input ∉ PI  then
14                  SIG_temp = SIG_temp ∪ input
15      }
16      SIG = SIG_temp
17      time = time + 1
18  }
19  end  algorithm
```

The underlying idea is that a signal s_t represents a signal s of the original circuit at time step t. In the TAM engine, first we convert the delays of the original gates and wires into the functional domain according to the chosen time unit. An *original gate* g with delay n is converted into n successive *untimed gates*: $(g, Buf_{n−1}, \ldots, Buf_1)$. These untimed gates model n time steps required to propagate a value from the input of g to the output of g (Fig. 6.3a). The equivalent untimed gates show the behavior of the original gate with an accuracy of one time unit. Also, a wire w with delay n is converted to n successive buffers: $(Buf_n, \ldots, Buf_1)$ (Fig. 6.3b). The circuit with the untimed gates and wires is called the *untimed circuit*. Untimed gates are also used for ATPG of crosstalk faults in [GK10].

Having the untimed circuit, the second step of the TAM engine begins to convert the timing behavior of the overall circuit into the functional domain. The TAM algorithm is described in Fig. 6.4 in pseudo code. The input data of this algorithm is an untimed circuit. For each gate the algorithm creates as many copies as values of the gate at different time steps may be relevant to determine the output.

The algorithm starts from *Primary Outputs* (PO) and traverses the untimed circuit graph backward (line 3). Given the delay of one time unit at each gate in the untimed circuit, the output of an untimed gate depends on its corresponding inputs one time

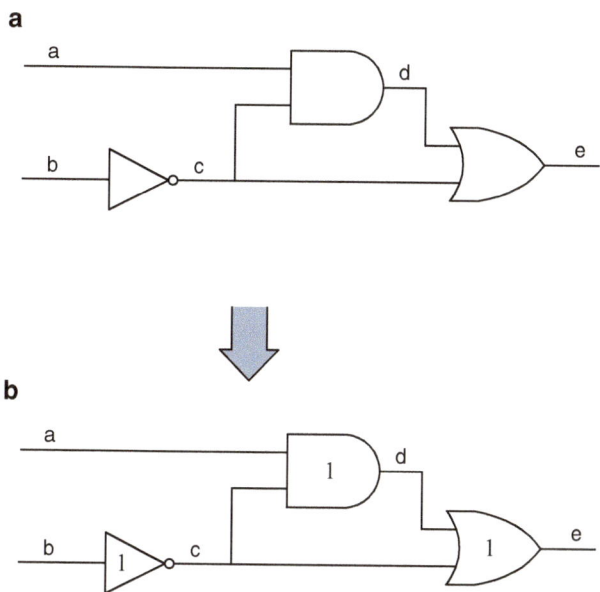

Fig. 6.5 (a) Original circuit and original gates (b) Untimed circuit and untimed gates

step ago. These inputs are given by the predecessor node of the output in the untimed circuit graph (line 9). Given the output and its driving inputs, a new gate is created in which the output and the inputs have the timing difference of one time unit (line 10). The newly created gates are called *TAM gates*. A new circuit called the *TAM circuit* is constituted by the TAM gates. The inputs of the current untimed gates are collected in the set *SIG_temp* to be used for the next backward traversal step (lines 12–14). If the input is a *Primary Input* (PI) or already exists in the set *SIG_temp* (fanout case), it will not be added to the set *SIG_temp* (line 13).

Example 6.1. We explain the TAM algorithm using the example original circuit of Fig. 6.5a. For the sake of simplicity, the delays of the original gates are considered to be 1. Therefore, the untimed circuit (Fig. 6.5b) is the same as the original circuit. In the first step, the algorithm starts from primary output e and copies the OR gate: $OR(c_{t-1}, d_{t-1}, e_t)$. Figure 6.6 shows the copied gates (TAM gates). The dotted lines visualize the time steps. $SIG\_temp = \{c, d\}$ is created. In the second step, the untimed circuit is traversed from c and d backwards. The NOT and AND gates are copied: $NOT(b_{t-2}, c_{t-1}), AND(a_{t-2}, c_{t-2}, d_{t-1})$. Having $SIG\_temp = \{c\}$, the third step starts and copies the NOT gate: $NOT(b_{t-3}, c_{t-2})$. As explained in Sect. 6.1, O_t depends on the inputs of $I_{t-D_s}, I_{t-D_s-1}, \ldots, I_{t-D_l}$ where $D_s = 2$ and $D_l = 3$ in this example.

For investigating the behavior of a circuit under timing variations, the VL is inserted in the TAM. The slowdown of a signal by d time units is modeled by the VL. The slowdown is modeled for each gate independently by activating the

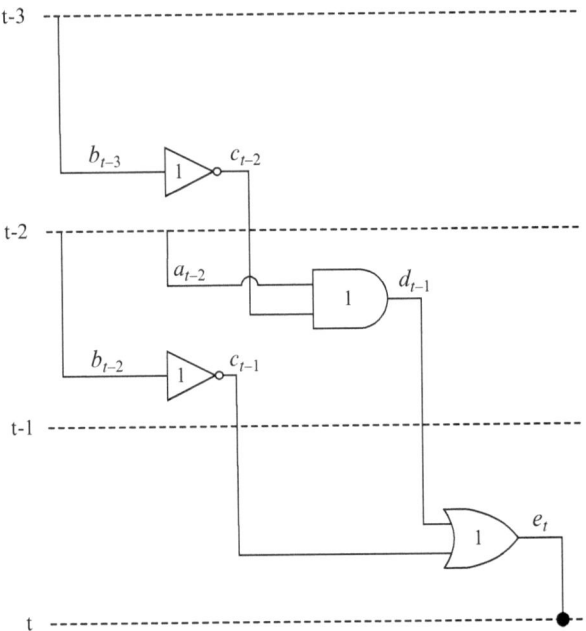

Fig. 6.6 TAM circuit and TAM gates

select line of the variation multiplexer of the corresponding gate. Therefore, in
addition to the value of a signal in the corresponding time, the value of the signal d
time steps ago is also needed. We describe this algorithm in Fig. 6.7 which extends
the algorithm of Fig. 6.4.

In lines 10–11 of Fig. 6.7, when $d = 0$, the gate related to the corresponding
time is copied. This gate is called TAM gate. After that, the behavior of the gate
up to d time steps ago is also copied (lines 10–11, when $d \neq 0$). These gates are
called *slowdown gates*. A maximum timing variation D specified by the user. This
maximum timing variation is used to limit parameter d. Given a TAM gate ($gate_0$)
and its corresponding slowdown gates ($gate_1, \cdots, gate_D$), a variation multiplexer is
added (lines 14–15). The variation multiplexer can skew the normal behavior of a
gate by a slowdown up to D time units. The timing variation of each original gate is
modeled by its corresponding variation multiplexers (line 13). Additional slowdown
in the input cone of slowdown gates may accumulate to an overall slowdown up to
D time units (lines 14–15, when $d \neq 0$).

The circuit generated by the algorithm is shown in Fig. 6.8 where $D = 1$. The
dashed gates indicate the slowdown gates. For each original gate, one variation
multiplexer is inserted at the output of its corresponding TAM gate. A variation
multiplexer either selects the normal behavior of a signal or the signal value one
time step ago.

```
 1  algorithm  TAM+VL (In : untimed circuit, Out : TAM+VL circuit)
 2  time = 0
 3  SIG = PO
 4  while  SIG ≠ ∅  do
 5  {
 6     SIG_temp = ∅
 7     foreach  sig ∈ SIG  do
 8     {
 9        gate = predecessor(sig)
10        for  0 ≤ d ≤ D  do
11           copy(gate_d, i_{t-time-1-d}, o_{t-time-d})
12
13        if  sig is Original_Gate_Output  then
14           for  0 ≤ d < D  do
15              Insert_Variation_Mux_d(in : gate_d, ⋯, gate_D)
16
17        foreach  input ∈ I(gate)  do
18           if  input ∉ SIG_temp  and  input ∉ PI  then
19              SIG_temp = SIG_temp ∪ input
20     }
21     SIG = SIG_temp
22     time = time + 1
23  }
24  end algorithm
```

Fig. 6.7 TAM + VL

Overall, the generated circuit is formulated as follows:

$$\Phi = C(IN(t)) \cdot \prod_{d=1}^{D} C^d(IN(t-d)) \cdot M \tag{6.2}$$

$$IN(t) :\ I_{t-D_s},\ I_{t-D_s-1},\ \ldots,\ I_{t-D_l}$$

Φ has three main parts: TAM circuit (C), slowdown circuit (C^d), and variation multiplexers (M). The parts M and C^d together are called variation logic.

The topology and the delay of reconverging paths of the original circuit affect the size of the TAM. To build the TAM, the circuit is traversed from the output backwards. If a fanout is visited whose different branches have reconverged with different delays, the input cone of the corresponding fanout is copied one time for each branch. In the worst case, the size of the TAM may be exponentially larger than the original circuit.

6.2.2   Time Control and Variation Control

Parameters f and T denote frequency and clock period, respectively. The relation of frequency and clock period is denoted by $f = 1/T$. The task of the TC is applying

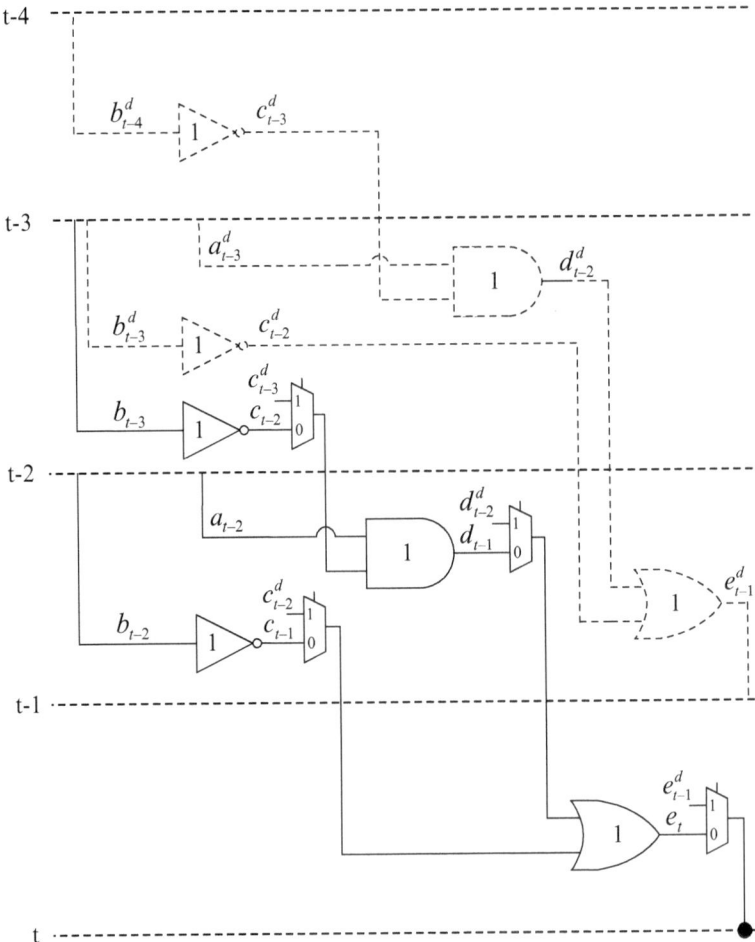

Fig. 6.8 TAM and VL

a clock period T on the inputs with the accuracy of one time unit. According to
Formula (6.1), inputs are constrained to have a constant value throughout each clock
cycle.

The task of the VC is controlling the select lines of the variation multiplexers.
The following constraint is used by the VC to apply the maximum timing variation:

$$\sum_{i=1}^{n} sel_i \leq D \qquad (6.3)$$

where sel_i denotes the integer value of the select lines related to one variation
multiplexer.

Fig. 6.9 Slowdown

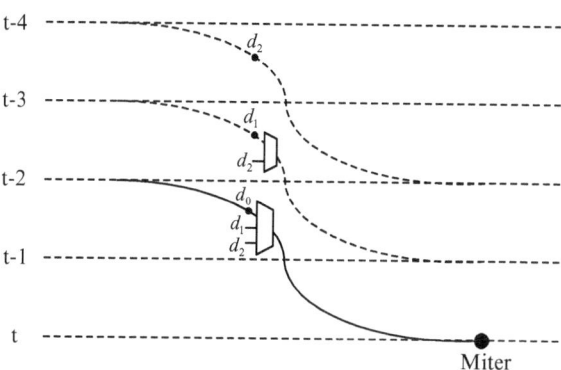

More complex variations can also be modeled by adding alternative constraints. For example to apply and to control block-based variation models [BCSS08], variations of each region can be controlled by a constraint. In a hierarchical manner, the variations of different regions in each level of hierarchy (like quadtree partitioning [ABZ03]) can be correlated by additional constraints in each level.

Clock skew can also be considered by extending our approach. This can be done by adding some units for the delay of the clock network and the sequential elements.

6.2.3 Slowdown Versus Speedup

In the previous sections, we showed how the slowdown induced under timing variations is modeled. Here, we discuss how to analyze a circuit under slowdown, speedup or both.

In the circuit, an original gate g has a specified delay n. The maximum *slowdown* and maximum *speedup* of an original gate g are considered to be y and x, respectively. The total *timing variation* is $D = x + y$. Therefore, the delay value of an original gate is bounded: $n - x \leq Gate\_Delay \leq n + y$.

When $x = 0$ and $y > 0$, the algorithm of Fig. 6.7 is invoked to apply the slowdowns. Figure 6.9 shows an example path and a gate with $x = 0$ and $y = 2$ such that the paths including the slowdown gates are denoted by d_1 and d_2. The variation multiplexer can skew the normal behavior (d_0) by selecting a slowdown of one time unit (d_1) or a slowdown of two time units (d_2). Also the variation multiplexer on the delay path d_1 can skew the gate behavior by another time unit such that the overall applied skew is at most 2. Here, the reference time is t and the values of primary inputs used for previous times ($t - 1$, $t - 2$, ...) is controlled by the TC.

When the effect of speedup ($x > 0$ and $y = 0$) is evaluated, again the algorithm of Fig. 6.7 can be reused with a minor modification. First, the position of the variation multiplexer changes. Figure 6.10 shows this case by an example when

Fig. 6.10 Speedup

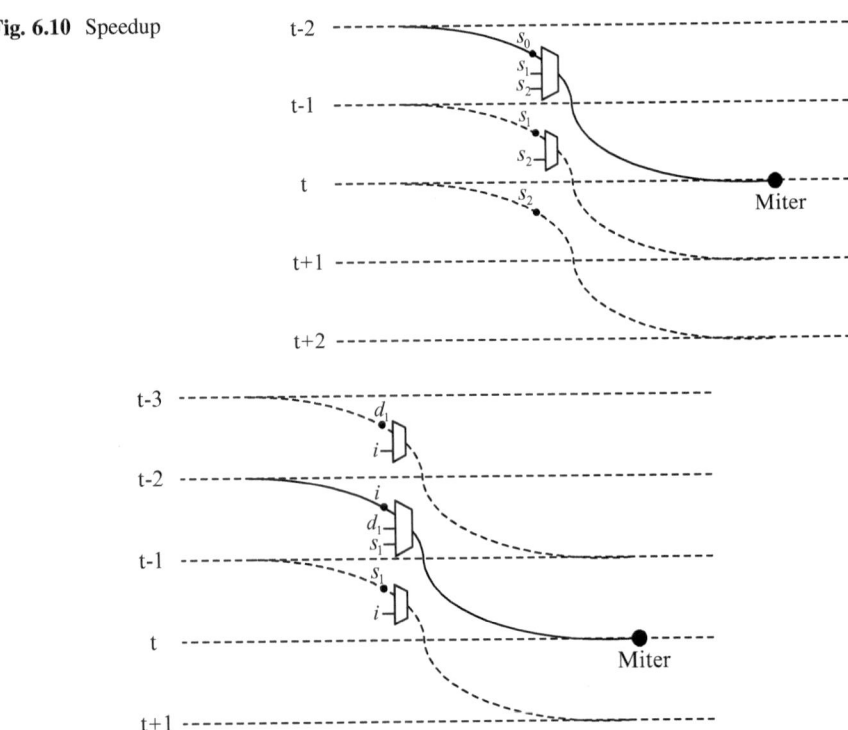

Fig. 6.11 Slowdown and speedup

Fig. 6.12 Transition
monitoring

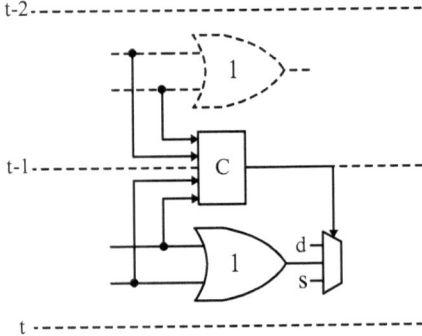

$x = 2$ and $y = 0$. Now the variation multiplexer can skew the normal behavior
(s_0) by selecting a speedup of one time unit (s_1) or a speedup of two time units (s_2).
Also the variation multiplexer on the speedup path s_1 can skew the gate behavior by
another one time unit such that the overall applied skew is at most 2. Here, also the
position of the reference time t changes. In this case, the TC controls the frequency
of the previous times ($t-1, t-2, \dots$) as well as the next times ($t, t+1, t+2, \dots$).

In the case of $x > 0$ and $y > 0$, there are both slowdown and speedup. In this case, the algorithm of Fig. 6.7 is invoked with the input $D = x + y$. For example, when $x = 1$ and $y = 1$, the algorithm of Fig. 6.7 is invoked with $D = 1 + 1 = 2$. Here, the position of the variation multiplexer as well as the reference time t change. This example is shown by Fig. 6.11. In this case, the variation multiplexer can skew the normal behavior (i) by a delay of one time unit (d_1) or by a speedup of one time unit (s_1). Also the variation multiplexer on the delay path d_1 can skew the gate behavior by a speedup of one time unit (i). The variation multiplexer on the speedup path s_1 can skew the gate behavior by a delay of one time unit (i).

More complex delay models can also be handled using VC. For modeling the behavior of a gate where the gate delay depends on the transitions of its inputs, we require additional constraints to monitor the input's transitions and to activate the corresponding slowdown or speedup. In this case again, we consider each gate having a specified delay. But this delay may increase or decrease dependent on the input's transitions. To monitor a transition, the behavior of each gate during two consecutive time steps is required. Therefore, in addition to the normal behavior of a gate, another copy of the gate one time step ago is created. As Fig. 6.12 shows, one additional gate is copied. Then, the *transition constraint C* activates the select lines of the multiplexer (slowdown or speedup) dependent on the transitions.

6.3   Properties of TAM

In this section, properties of the TAM are discussed and proved. First the properties of the TAM for the outputs at one time step are shown. Then we show how waveforms are modeled using the TAM.

Definition 6.1. The *Arrival Time* (AT) of a signal at the output of a gate p is the time required for the value of the signal to propagate from PIs to the output of the gate p. This time is the sum of delays of gates along a path. The maximum arrival time AT_{max} at the output of a gate p is the maximum amount of the time which the value of a signal requires to propagate from PIs to the output of the gate p. The minimum arrival time AT_{min} at the output of a gate p is the minimum amount of the time which the value of a signal requires to propagate from PIs to the output of the gate p [MB91, SCPB12].

Definition 6.2. The *Propagation Time* (PT_p^P) of a signal at the output of a gate p along a particular path P is the time required for the value of the signal at the output of the gate p to propagate to the corresponding PO [VARR11].

Definition 6.3. The value of a signal at the output of a gate p along a path P at time step $t - PT_p^P$ affects the value of the corresponding PO at time step t.

The definitions in this section are based on the timing model presented in Sect. 6.2. However, for the sake of simplicity, we assume that values of the PIs change at the same time [SCPB12]. The reference time is the time step t which is the time step of the output sampling. Other time steps are relative times to the reference time. In our timing model, pattern-dependent delay models [TBG11] and correlation of the timing parameters [BCSS08] are not considered.

Definition 6.4. The model TAM_t is a TAM model which considers the output o at time step t (o_t).

Theorem 6.1. *The TAM_t models all time steps of internal signals and inputs which affect the output o at time step t (o_t).*

Proof. The TAM model is constructed by traversing an untimed circuit stepwise via *Breadth-First Search* (BFS) from the output o towards PIs. The algorithm visits every gate at least once. If a gate p is visited, three cases may occur as follows:

- Case 1: The gate p is visited only once. In this case, the signal s of the output of the gate p never reconverges. Because if the signal s reconverges, the gate p is visited at least twice while traversing the circuit by the algorithm from the output o backwards. Therefore, the value of the signal s only at one time step affects the value of the output o at the sampled time step t.
- Case 2: The gate p is visited through paths with the same PTs. In this case, only the value of the signal s at one time step affects the output o through multiple paths. Because the paths have the same PTs to the output o. Line 13 (the *if*-expression) in the algorithm of Fig. 6.4 checks whether a gate has been visited through paths with the same PT. If a gate is visited for the first time through a path, the *if*-expression becomes true. But if a gate is visited for the second time through a path with the same PT, the *if*-expression becomes false and the corresponding gate only once at one time step is copied.
- Case 3: The gate p is visited through paths with different PTs. In this case, the value of the signal s at different time steps affects the output o at the sampled time step t. In this case, the *if*-expression of line 13 in the algorithm is false every time. Therefore, the gate p is copied for all corresponding time steps.

The model TAM_t is constructed with respect to the time step of the output sampling. However, more sampling time steps are required to observe the behavior of a waveform on the output, i.e., possible transitions and glitches on the output. When minimum and maximum arrival times of the output o are AT_{min} and AT_{max}, i.e., $AT_{min}(o)$ and $AT_{max}(o)$, all possible transitions and glitches on the output o happen in the time interval $[AT_{min}(o), \ldots, AT_{max}(o)]$. The *waveform* of the output o is $[o_u, \ldots, o_v]$ where $u = AT_{min}(o)$ and $v = AT_{max}(o)$ and each o_t ($u \leq t \leq v$) is the value of the output o at the time step t. For simulating a test vector (I_0, I_1), all inputs are assumed to switch their values simultaneously at one time step [SCPB12]. We assume that each wire has stabilized its initial value under I_0 before the switching time of the inputs [SCPB12]. The peaks (transitions) are propagated through the internal signals and arrive at the output. The reconvergences create additional transitions and glitches on signals. Before the time step $AT_{min}(o)$, the output o has an initial value. After the time step $AT_{max}(o)$, the output o has a stable value. The waveform of the output of the gate p is in the time interval $[AT_{min}(p), \ldots, AT_{max}(p)]$, where $AT_{min}(p) \leq AT_{min}(o)$ and $AT_{max}(p) \leq AT_{max}(o)$.

Theorem 6.2. *The waveform* $[p_k, \ldots, p_l]$ *of the output of the gate* p, *where* $k = AT_{min}(p)$ *and* $l = AT_{max}(p)$, *is a subinterval of* $[p_m, \ldots, p_n]$, *where* $m = AT_{min}(o) - PT_p^P$ *and* $n = AT_{max}(o) - PT_p^P$, *i.e.,* $[p_k, \ldots, p_l] \subseteq [p_m, \ldots, p_n]$

Proof. The expression $AT_{min}(o) \leq AT_{min}(p) + PT_p^P$ is true because if $AT_{min}(p) + PT_p^P$ corresponds to the shortest path in the circuit, then $AT_{min}(o) = AT_{min}(p) + PT_p^P$, otherwise $AT_{min}(o) < AT_{min}(p) + PT_p^P$. Also, $AT_{max}(o) \geq AT_{max}(p) + PT_p^P$ because if $AT_{max}(p) + PT_p^P$ corresponds to the longest path in the circuit, then $AT_{max}(o) = AT_{max}(p) + PT_p^P$, otherwise $AT_{max}(o) > AT_{max}(p) + PT_p^P$. Therefore, we have the following expressions:

$$\begin{cases} AT_{min}(o) \leq AT_{min}(p) + PT_p^P \\ AT_{max}(o) \geq AT_{max}(p) + PT_p^P \end{cases} \tag{6.4}$$

which results in:

$$\begin{cases} AT_{min}(o) - PT_p^P \leq AT_{min}(p) \\ AT_{max}(o) - PT_p^P \geq AT_{max}(p) \end{cases} \tag{6.5}$$

Formula (6.5) implies that the time interval $[AT_{min}(p), \ldots, AT_{max}(p)]$ is a subinterval of $[AT_{min}(o) - PT_p^P, \ldots, AT_{max}(o) - PT_p^P]$. Thus, the waveform $[p_k, \ldots, p_l]$ is a subwaveform of $[p_m, \ldots, p_n]$.

Lemma 6.1. *Let* P *be a path of an untimed circuit and* p *be the output of a gate along the path* P, *then a signal for the output of the gate* p *in* TAM_t *is created by the algorithm of Fig. 6.4 at time step* $t - PT_p^P$.

Proof. The lemma is true by the construction of the algorithm in Fig. 6.4. The algorithm creates the TAM_t in a BFS. The algorithm starts from the output o at the time step t. In each iteration, all gates reached from POs over paths with PT time units are visited. In this case, the corresponding gates are copied for the time step $t - PT$.

Multiple TAMs can be constructed to model the behavior of the waveform of the output o. In this case, $TAM_i, TAM_{i+1}, \ldots, TAM_j$, where $i = AT_{min}$ and $j = AT_{max}$, are created. The algorithm of Fig. 6.4 is adapted to create multiple TAMs. In the first iteration of the algorithm, an untimed circuit is traversed backwards from the output o at time step $t = j$ creating TAM_j. In the next iteration, the signals involved at time step $t = j - 1$ are traversed backwards. In this step, also the algorithm has to traverse again the output o at time step $j - 1$ backwards to create TAM_{j-1}. In this case, not only the signals in the set SIG_temp but also a new signal for the output o is added to the set SIG in line 16 of the algorithm of Fig. 6.4. Then, the signals in the set *SIG* are traversed backwards. This procedure repeats until all TAMs have been created.

Theorem 6.3. *If the minimum and the maximum arrival times of the circuit output o are AT_{min} and AT_{max}, then $TAM_i, TAM_{i+1}, \ldots, TAM_j$, where $i = AT_{min}$ and $j = AT_{max}$, model all possible transitions and glitches of internal signals and outputs created by transitions of PIs.*

Proof. $TAM_i, \ldots, TAM_j$ contain all the time steps of the output o (output waveform) in the time interval $[AT_{min}(o), \ldots, AT_{max}(o)]$. Also the copies $TAM_i, \ldots, TAM_j$ of the circuit contain all time steps of the output of the gate p in the time interval $[AT_{min}(o) - PT_p^P, \ldots, AT_{max}(o) - PT_p^P]$ because TAM_t contains a copy of the gate p at time step $t - PT_p^P$ (Lemma 1). To build TAM_t the algorithm starts from the output o and traverses the circuit back. When it arrives at the output of the gate p, a signal for the output of this gate at time step $t - PT_p^P$ is copied. Also, the waveform of the gate output is in the time interval $[AT_{min}(o) - PT_p^P, \ldots, AT_{max}(o) - PT_p^P]$ as proved in Theorem 2. Therefore, $TAM_i, \ldots, TAM_j$ model the waveform (all possible transitions and glitches) of the output of the gate p.

The effect of all reconvergences regarding the output of each gate p is modeled using $TAM_i, \ldots, TAM_j$, where $i = AT_{min}$ and $j = AT_{max}$, because all glitches derived from reconvergences happen in the time interval $[AT_{min}(p), \ldots, AT_{max}(p)]$ for each gate p.

If there is a maximum timing variation D (x time units speedup and y time units slowdown), D additional copies of the TAM are required in order to model all possible transitions and glitches in the presence of timing variations. As explained in Sect. 6.2.3, in this case, multiplexers can activate a timing variation and can create a new waveform with respect to the activated timing variation.

The TAM model is used to analyze the timing behavior of a circuit under timing variations. The size of the TAM influences the time required to analyze the timing behavior of a circuit. Therefore, an upper bound to the size of the TAM can be used to predict the memory consumption and the time required to analyze circuits. In the following, we show how to determine an upper bound to the size of the TAM.

To construct the TAM, an untimed circuit is traversed from the output to the inputs. During the backward traversal, if a fan-out is visited which affects the value of the output at the sampled time step through its different branches at different time steps, the input cone of the corresponding fanout is copied one time for each affecting branch. Traversing from one output backwards, each reached fan-out indicates at least one prior reconvergence. In the worst case, all branches of each fanout may influence the value of the output at the sampled time step. Therefore, an upper bound to the size of the TAM can be predicted by analyzing the structure of a circuit. To calculate the upper bound to the size of the TAM, a circuit is traversed backwards. When a fan-out is reached, the input cone of the corresponding fan-out is copied one time for each branch. The section of experimental results will also show the upper bound to the size of the TAM for the benchmarks.

6.4 Experimental Results

The proposed approach is utilized to analyze arithmetic circuits under timing variations. The experiments are carried out on a Quad-Core AMD Phenom(tm) II X4 965 Processor (3.4 GHz, 8 GB main memory) running Linux. Our circuits are synthesized by Synopsys Design Compiler with Nangate 45nm Open Cell Library [Nan11]. The techniques described in this chapter are implemented using C++ in the WoLFram environment [SKF+09]. For the experiments, one time unit is 0.01 ns. MiniSAT is used as underlying SAT solver [ES04].

Ripple Carry Adder (RCA), *Carry Look-ahead Adder* (CLA), *Brent-Kung Adder* (BKA), and *Ladner-Fischer Adder* (LFA) benchmarks are evaluated by our approach against timing variations.

As Fig. 6.2 showed, our framework includes two side components: *spec* and *miter*. In the experiments, the original circuit is considered as a specification. The inputs of the most recent clock cycle are applied to the specification. For arithmetic circuits, we use a miter on the outputs of the specification and the TAM to measure the output deviation as the numerical difference. This miter is an integer subtractor followed by a comparison operation (Fig. 6.13). In order to measure the *maximum positive error*, the miter subtracts the specification output (O) from the TAM output (O_t): ($O_t - O \geq L$). To compute the *maximum negative error*, the following miter is used: ($O - O_t \geq L$).

We use a binary search to determine the maximum error. First an interval $[a,b)$ is selected such that when $L = a$, the CNF is satisfiable and when $L = b$, the CNF is unsatisfiable, i.e., the maximum error is more equal than a and less than b. In this case, the interval is divided into two smaller intervals by selecting $c = a + ((b-a)/2)$ as the middle point of the last interval. If the CNF is satisfiable with $L = c$, the next interval will be $[c,b)$, otherwise the next interval will be $[a,c)$. This procedure repeats until a sufficient accuracy is obtained. Considering $[a,b)$ as the final interval, The accuracy k is specified by the length of the final interval i divided by maximum 2^n (n = number of output bits), i.e, $k = i/2^n$. In this case, b is considered as an upper bound for the maximum error.

Figures 6.14 and 6.15 show the maximum positive and negative error computed for an 8-bit RCA and an 8-bit CLA under overclocking and in the presence of timing variations. The X-axis indicates the frequency f as the inverse of the clock

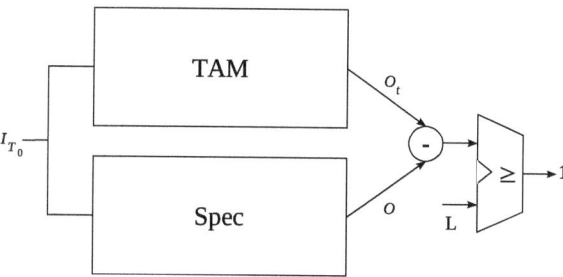

Fig. 6.13 Maximum error constraint

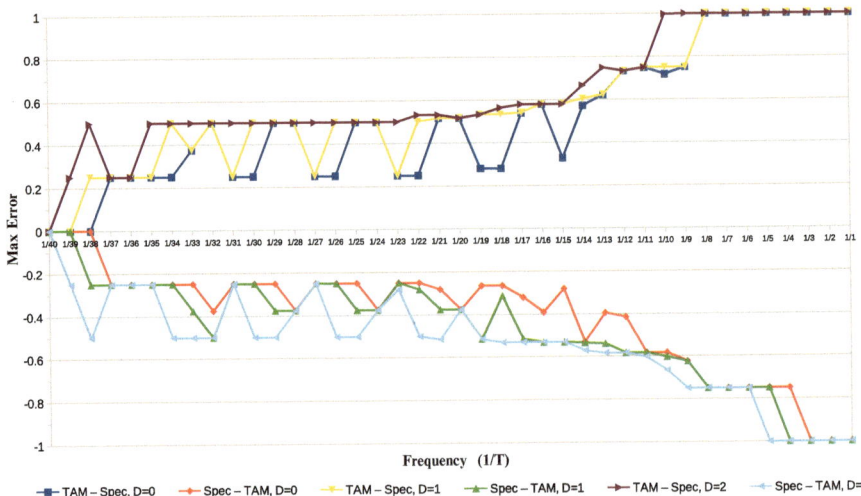

Fig. 6.14 Maximum error for 8-bit RCA

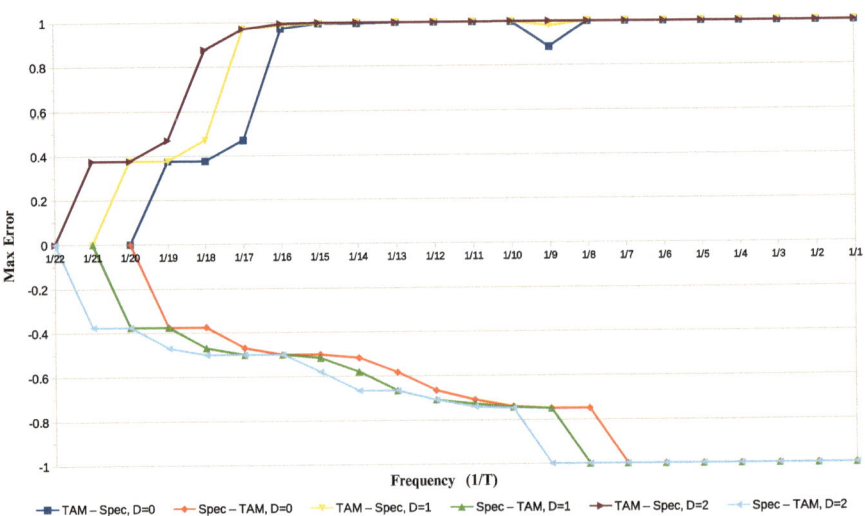

Fig. 6.15 Maximum error for 8-bit CLA

period T ($f = 1/T$). The minimum frequency specified by the X-axis is related
to the maximum delay of the circuit considering the maximum delay variation D.
The Y-axis indicates the maximum error as a result of the computed error divided
by the maximum output value. When $D = 0$, no timing variation is activated.
In this case, the diagram shows the maximum error caused by clock overscaling.
While the frequency increases, the maximum error sometimes decreases. This
is because the failing output bits are functionally correlated as they may share

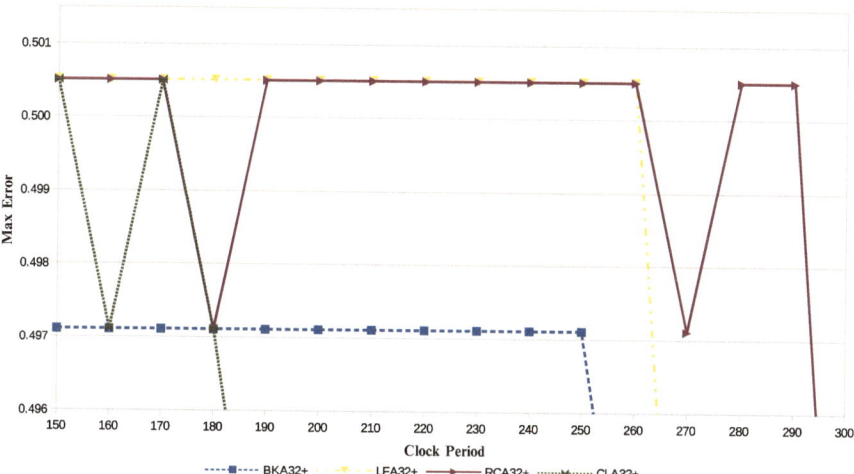

Fig. 6.16 Maximum error for 32-bit adders when D = 0

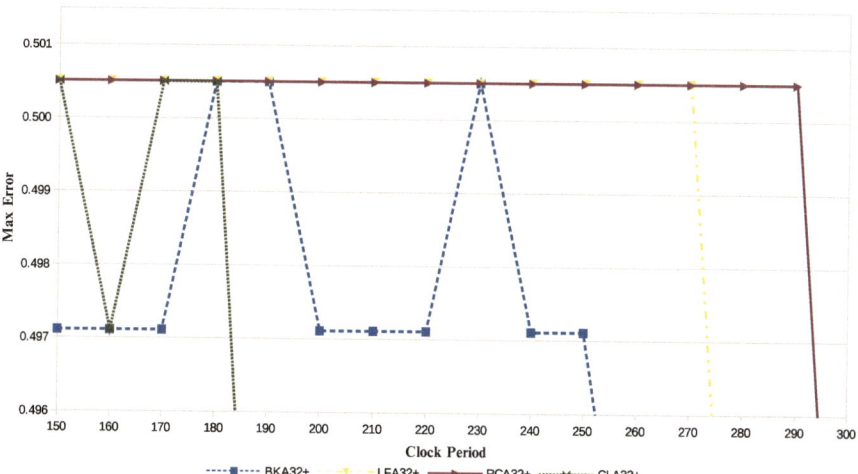

Fig. 6.17 Maximum error for 32-bit adders when D = 1

certain paths. Also, the maximum positive error and the maximum negative error may have different values at the same frequency because overscaling modifies the functionality and induces asymmetry in the circuit. When the frequency increases for the CLA, the maximum error increases faster (Fig. 6.15). This is due to the fact that more MSBs have a closer delay. Thus, multiple MSBs may fail together by overclocking. Therefore, the CLA is more sensitive to overclocking. But for the RCA, the maximum error increases slower than for the CLA because the MSBs have different delays. Therefore, they fail gradually along overclocking. In Fig. 6.14, the delay of the circuit is 38 time units. Therefore when $D = 0$ and $f = 1/38$, there is

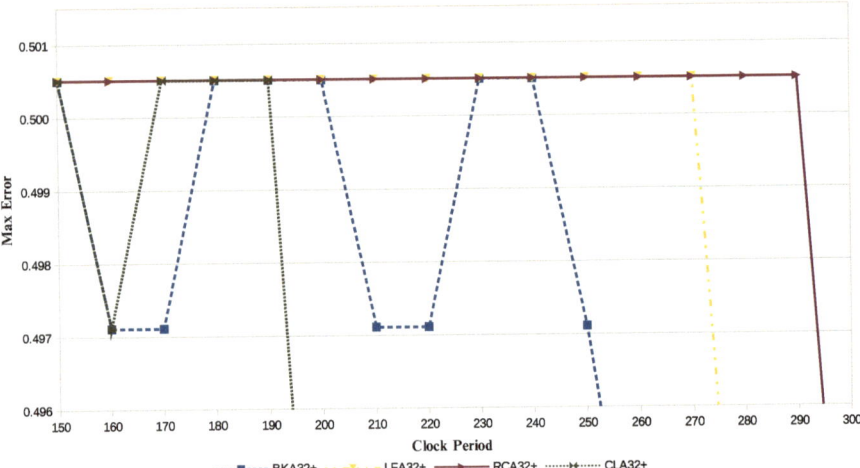

Fig. 6.18 Maximum error for 32-bit adders when D = 2

no error at the output of the circuit. When the clock is overscaled, for example the frequency increases from $f = 1/38$ to $f = 1/37$, then the maximum error 0.25 is observed on the outputs.

The maximum positive error computed for 32-bit adders when $D = 0$, $D = 1$, and $D = 2$ is shown in Figs. 6.16, 6.17, and 6.18. For 32-bit adders, the accuracy is $k = 0.0007$ and the maximum observed error in the diagrams is 0.50001. Among the 32-bit adders, CLA has the shortest delay. After CLA; BKA, LFA, and RCA have the shortest delay. The maximum error computed along overclocking is shown in Fig. 6.16. In the determined clock periods, the maximum error of LFA is 0.500, while the maximum error of BKA is 0.497. At $T = 270$ and $T = 180$, the maximum error of RCA decreases. For CLA, at $T = 180$ and $T = 160$, we observe a decreased error. While the clock period decreases, sometimes the maximum error also decreases. This is due to failing output bits that are functionally correlated as they may share certain paths.

The timing variation is activated ($D = 1$) in Fig. 6.17. At the clock period $T = 180$ in Fig. 6.17, CLA maximum error increases in comparison to the non-varied CLA of Fig. 6.16. When $D = 1$, RCA maximum error increases at the clock period $T = 270$ and $T = 180$. But the errors of LFA do not increase. For BKA, at the clock period $T = 230$, $T = 190$, and $T = 180$, an increased error is observed.

When $D = 2$ (Fig. 6.18), at the clock period $T = 190$, the CLA error increases. Also, at the clock period $T = 240$, $T = 200$, and $T = 150$, the BKA error increases. In this case, the maximum errors of RCA and LFA do not change in comparison to their errors in Fig. 6.17 (when $D = 1$). There is always an increase if there is at least one path "about to become critical". But this may not always be the case and we have a measurement error of $k = 0.0007$. Therefore, if an error increase is less than k, it is not visible in the results. As the diagrams showed, the

Table 6.1 Size and time

Circ Name	Size when Time Unit = 0.01 ns				Time when D = 2
	Original Circ	Untimed Circ	TAM Circ	Upper Bound	Time (s)
BKA32	136	637	7561	8109	747
LFA32	135	634	7982	8074	1428
RCA32	128	608	7712	7712	868
CLA32	153	710	9154	10910	1402

used adders are not resilient against overclocking and timing variations. Other kinds of approximate circuits can be utilized to bound the error [MCRR11,GMP+11] and consequently to improve the robustness of the circuit.

Table 6.1 presents the number of gates in the original circuit, the untimed circuit, the TAM circuit, and the upper bound to the size of the TAM as well as the average run times required to compute the maximum error at one clock period for 32-bit adders. The number of original gates for BKA is 136, while the number of untimed gates is 637. It indicates that 501 (637–136) additional buffers have been added to the original circuit to create the untimed circuit. CLA has the largest number of TAM gates. This shows that in CLA there are more paths than in others which consequently increase the size of the TAM.

6.5 Summary

In this chapter, we introduced an approach to model and to analyze the functional behavior of circuits under timing variations. The framework includes the following main components: TAM and VL, TC, and VC. Our framework is utilized to analyze a circuit under timing variations as well as an approximate circuit under timing-induced errors.

The TAM and timing variation model presented in this chapter are utilized to automate speedpath debugging in the next chapter. As the TAM and timing variation model are translated into the CNF, an enhanced SAT-based debugging is proposed in Chap. 7 to automate debugging of the circuits under timing variations.

Chapter 7
Automated Debugging for Timing Variations

This chapter deals with the automation of post-silicon debugging for speed-limiting paths, briefly called *speedpaths*. Debugging of speedpaths is a key challenge in development of VLSI circuits as timing variations induced by process and environmental effects are increasing.

A *speedpath* is a frequency-limiting critical path which affects the performance of a chip [BKWC08, LWPM05]. A speedpath that violates timing constraints at the post-silicon stage is called *failing speedpath* [XDS10]. Speedpaths fail due to, e.g., timing variations induced by process, design and environmental effects [BKWC08].

The correct behavior of a chip is validated at the post-silicon stage by applying test vectors to the chip. When a speed failure is detected due to violating frequency constraints [KNKB08], the debug team identifies failing speedpaths. But this is a time-consuming process which requires a large effort. Thus, automated debugging approaches to identify failing speedpaths are necessary to speed up the process.

A formal procedure based on an *Integer Linear Programming* (ILP) formulation is proposed in [XDS10] to diagnose segments of failing speedpaths due to process variations. The approach identifies segments of failing speedpaths that have a post-silicon delay larger than their estimated delay at the pre-silicon stage. *Parameterized Static Timing Analysis* (PSTA) is used in [OHN09] to obtain a variational model for every candidate speedpath from a given set of potential candidates. These variational models are then combined to create a cost function. This PSTA-based cost function is utilized by a branch and bound approach to determine the most probable failing speedpaths. The approach needs a set of user-supplied paths as inputs, and relies on post-silicon delay measurements prior to identifying failing speedpaths.

Clock shrinking on a tester combined with a CAD methodology is employed by the approach in [KNKB08] to isolate failing speedpaths. The work in [ZGC+10] presents a scan-based debug technique to failing speedpaths. The technique is based on at-speed scan test patterns. A trace-based approach is presented in [LX10] to debug failing speedpaths. The approach uses trace buffers to provide real-time visibility to the speedpaths during the normal operation. On-chip delay sensors are

© Springer International Publishing Switzerland 2015
M. Dehbashi, G. Fey, *Debug Automation from Pre-Silicon to Post-Silicon*,
DOI 10.1007/978–3–319-09309-3_7

used in [LDX12] to improve timing prediction and to utilize them in order to isolate failing speedpaths. Each sensor is a sequence of logic gates with an approximate location on the layout. In [MMSTR09], the diagnosis resolution is enhanced by processing failure logs at various slower-than-nominal clock frequencies. The work in [BKWC08] uses a statistical learning-based approach to predict failing speedpaths by measuring delays of a small set of representative speedpaths. A formal approach to find a small set of representative speedpaths in order to predict the timings of a large pool of target paths is proposed in [XD10]. A technique to automate debugging is presented in [MVL09] by using failing functional tests and then algorithmically isolating failing speedpaths. The technique uses functional implications without incorporating timing information. Using only functional implications limits the diagnosis accuracy. The SAT-based approach of [SVAV05] to automate debugging does not consider timing behavior of a circuit and timing variations. In Chap. 6, timing behavior of a circuit and timing variations were modeled in a functional domain.

In this chapter, we utilize this model to automate debugging of failing speedpaths [DF13a]. Given a circuit and an erroneous behavior observed on circuit outputs due to timing variations, potential failing speedpaths are automatically extracted by our approach. To automate debugging, first timing behavior of a circuit and corresponding timing variation models are converted into a functional domain as explained in Chap. 6. Then a debugging instance is formulated in CNF. Afterwards our algorithm extracts failing speedpaths using a SAT solver as an underlying engine. The diagnosis accuracy and the performance of the approach are experimentally shown on the ISCAS'85 and ISCAS'89 benchmark suites.

The remainder of this chapter is organized as follows. We introduce preliminary information on speedpath debugging in Sect. 7.1. Timing variation fault models are presented in Sect. 7.2. Then, our approach is presented in Sect. 7.3. The debugging algorithm is demonstrated in Sect. 7.4. Section 7.5 presents experimental results on benchmark circuits. Section 7.6 summarizes this chapter.

7.1 Speedpath Debugging

For the post-silicon validation, test vectors are applied to the chip and the clock period is reduced until an error is observed on outputs, registers or latches [OHN09, KKC07]. This step is called *clock shrinking*. The error is detected by comparing the output values of the chip with the nominal output values obtained from simulation at the specified clock period. The activating test vectors at the specified frequency and the observed error constitute an *Erroneous Trace* (ET). Having an erroneous trace, debugging starts to find failing speedpaths or some segments of failing speedpaths as fault candidates. In this chapter, a gate is considered as the smallest segment on a failing speedpath. Here, a fault candidate contains both spatial and temporal information about the sensitized gate.

Our assumption is that all registers are observable. In this case, an erroneous trace consists of input vectors of at least two clock cycles. An erroneous trace is denoted by ET and has the following parameters: $ET(I_0, I_1, O_2, T)$. Parameters I_0 and I_1 are the test vectors of two consecutive clock cycles causing an error. The observed error (erroneous output value) is shown by parameter O_2. Parameter T is the clock period with which the error was observed.

7.2 Fault Model

Slowdown and speedup fault models are presented in this section. These models are used in our experiments in order to inject faults and to evaluate a circuit against timing variations.

7.2.1 Slowdown Fault Model

An untimed AND gate with a delay of two time units is shown in Fig. 7.1a. To activate a slowdown of one time unit on a signal, the value of the signal one time step ago is required. Therefore, instead of selecting the value of a signal in the current time, its value at the previous time step is selected. To do this, a buffer with a delay of one time unit is inserted at the output of the untimed gate (Fig. 7.1b). In a simulation tool, the value of signal c_1 at one time step ago can be found on the successor signal c_2. A multiplexer is added to select c_1 or c_2. The multiplexer has zero delay. The select line of the multiplexer is controlled by signal sel. When $sel = 0$, signal c_1 is selected and the circuit has its normal behavior. If sel is activated for one time step, signal c_2 is selected which has the value of signal c_1 at one time step ago. The slowdown fault model can be used to evaluate the silicon effects like mis-modeled logic cells, capacitive-coupling, and voltage droop. These silicon effects are investigated in [OHN09].

7.2.2 Speedup Fault Model

A model to activate speedup is illustrated in Fig. 7.1c. It is assumed that a gate has a delay of more than one time unit. Therefore, there is at least one buffer at the output of an untimed gate. In this case, to activate a speedup of one time unit, no additional buffer is required. In Fig. 7.1c, signal c_2 holds the current value of the original gate output. The value of signal c_2 at the next time step can be found on signal c_1. The gate has a normal behavior if $sel = 0$. When $sel = 1$, a speedup of one time unit is activated by selecting signal c_1.

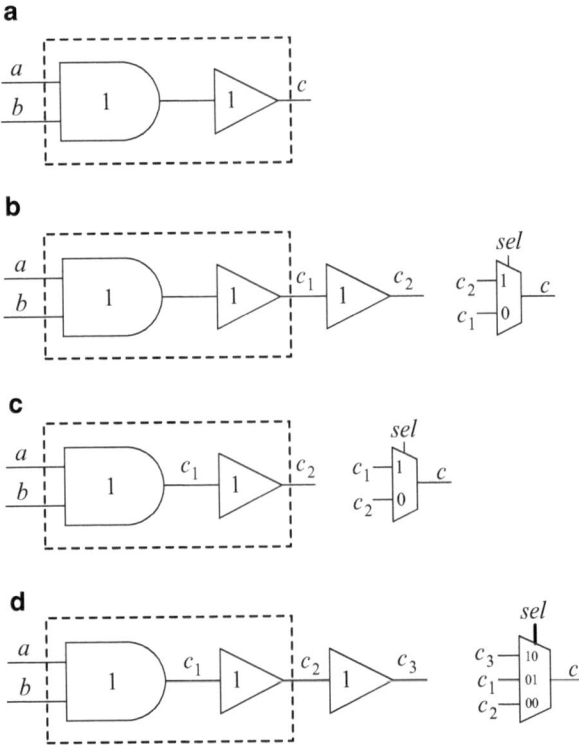

Fig. 7.1 Fault models for timing variations. (**a**) An untimed AND gate with delay $= 2$. (**b**) One-time-unit slowdown fault injection. (**c**) One-time-unit speedup fault injection. (**d**) One-time-unit slowdown/speedup fault injection

7.2.3 Slowdown and Speedup Fault Model

A multiplexer with three data inputs and one additional buffer is required to activate a speedup or a slowdown of one time unit. This case is shown in Fig. 7.1d. When $sel = 00$, normal behavior is selected (signal c_2). When $sel = 01$, a speedup of one time unit is activated by selecting signal c_1. When $sel = 10$, a slowdown of one time unit is activated by selecting signal c_3.

When signal sel is activated for one time step, a transient fault is injected. Signal sel can be activated permanently or with a special time distribution to inject a permanent fault or a distributed fault.

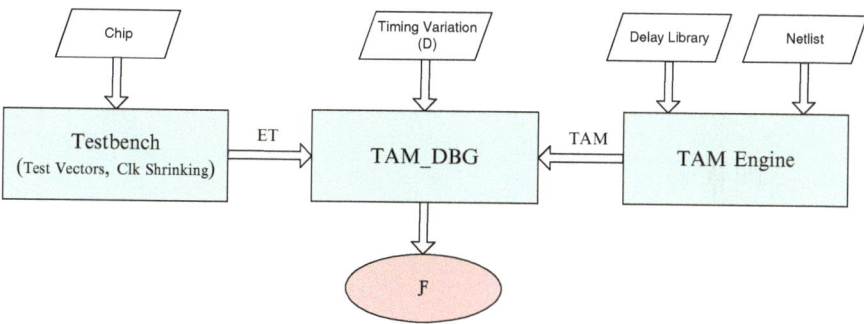

Fig. 7.2 Overview of proposed approach

7.3   Approach

The overall view of our approach is shown in Fig. 7.2. At the post-silicon stage, the correct timing behavior of a circuit is validated by applying test vectors to the chip while clock shrinking is performed. In Fig. 7.2, this step is performed in a testbench environment. When an erroneous trace is observed due to violating frequency constraints, debugging starts. Debugging automatically finds potential failing speedpaths. An ET consists of the test vectors activating a timing fault at a specified frequency and the corresponding erroneous output values.

Because the test vectors in consecutive clock cycles may activate a timing fault, test vectors should be applied at-speed to the chip. A tester can be used to apply test vectors unintrusively [KNKB08]. In microprocessor-based systems, SBST methods [PGSR10] can be effectively used to apply test vectors at-speed in order to validate the timing behavior of an internal module.

Having an erroneous trace, our goal is to automatically find potential speedpaths which have failed and have created the erroneous output values of the corresponding erroneous trace. To automate debugging, first the timing behavior of a circuit is converted into the functional domain as explained in Chap. 6. When the timing behavior of a circuit is available in the functional domain, formal verification methods can comprehensively analyze the timing effects of the circuit. In Fig. 7.2, the TAM engine models the timing behavior of a circuit in the functional domain with a discrete unit of delay. A netlist and a delay library are the inputs of the TAM engine. The output of the TAM engine is a time accurate model of the circuit, i.e., the TAM.

Having the TAM and an ET, the debugging process starts. TAM_DBG in Fig. 7.2 denotes this step. In the TAM_DBG engine, timing variation models are added according to the user-defined maximum timing variation D. The timing variation model can vary the value of a signal along the time axis. The behavior of timing variations is controlled by a *Variation Control* (VC) constraint. In this case, debugging investigates whether a timing variation on a signal is observable as the erroneous output value of the corresponding erroneous trace. This investigation

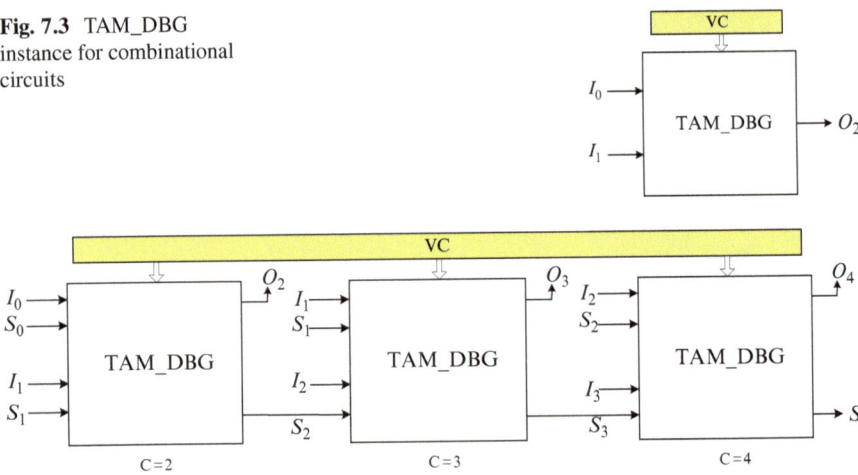

Fig. 7.3 TAM_DBG instance for combinational circuits

Fig. 7.4 TAM_DBG instance for sequential circuits

is performed by constraining the output and inputs of the created model to the values of the erroneous trace. Fault candidates whose timing variation may cause the erroneous behavior of the ET are identified by the TAM_DBG engine.

Our assumption is that combinational circuits are a part of synchronous circuits. Thus, the inputs to the combinational logic changes only once every cycle. For combinational circuits, the instance of Fig. 7.3 is created by our framework.

In sequential circuits, if all registers are observable, then they are treated like the combinational circuit of Fig. 7.3. In this case, I consists of both inputs and state bits (flipflops values). But when not all registers are observable, the error may be detected several clock cycles after fault activation. In this case, the sequential circuit is unrolled as many times as the number of clock cycles constituting the erroneous trace. In each unrolled clock cycle, a TAM_DBG model is created and the timing variations on the whole created instance is controlled by the VC. A sequential circuit is shown in Fig. 7.4 created by our approach. The erroneous trace has three clock cycles. In each clock cycle, a TAM_DBG model is used where its output (O_{i+1}, S_{i+1}) depends on the inputs and state bits of two clock cycles $(I_{i-1}, S_{i-1}, I_i, S_i)$. The overall model is formulated as follows:

$$\Phi_C = \prod_{i=1}^{n} TAM\_DBG\ (\ I_{i-1}, S_{i-1}, I_i, S_i,\ O_{i+1}, S_{i+1}\ ) \qquad (7.1)$$

Parameter n is the length of the erroneous trace which indicates the number of clock cycles needed to observe the error. The diagnosis accuracy depends on the granularity of the time unit and the accuracy of the variation models. A time unit has to be selected such that timing variations are an integer multiple of one time unit.

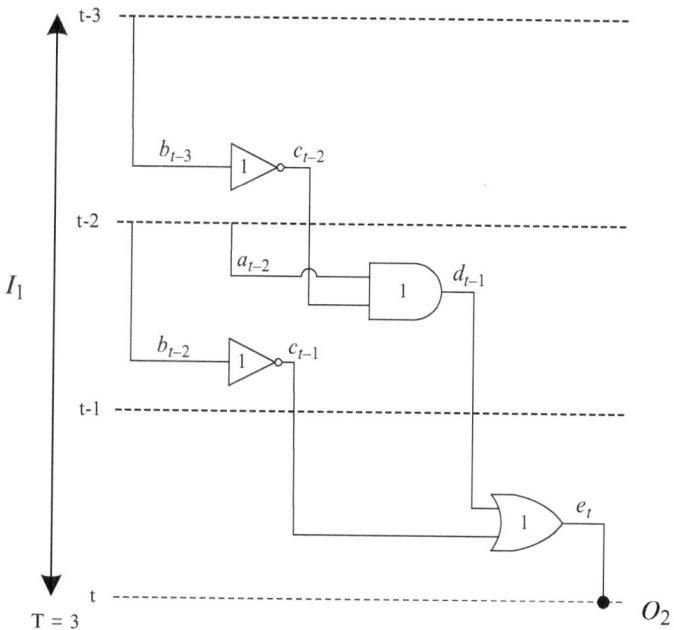

Fig. 7.5 TAM and TC when T = 3

7.4 TAM-Based Debugging

For debugging of failing speedpaths, first the TAM of the circuit is constructed as explained in Chap. 6. In the TAM circuit of Fig. 7.5, each signal s_t represents an original signal s at time step t. Furthermore, the TC constraint to model a clock period is applied to control the input values in different time steps. In Fig. 7.5, the clock period is $T = 3$. Therefore, the signals b_{t-2} and b_{t-3} have to have the same value.

D additional copies of the TAM are created in order to model maximum timing variation D. The case of $D = 1$ is shown in Fig. 7.6. Multiplexers on the outputs of TAM gates can activate a timing variation by selecting the value of a signal from different time steps. The select lines of multiplexers are controlled by VC modeling timing variations. When there is a maximum timing variation $D = 1$ at $T = 3$, the output O_2 may depend on the inputs of two clock cycles I_0 and I_1. In this case, the older inputs (I_0) affect the output through longer paths while the newer inputs (I_1) affect the output through shorter paths.

Having the instance of Fig. 7.6, debugging begins. The inputs I_0 and I_1 and the output O_2 are constrained to the values of the erroneous trace obtained from a testbench. Then, debugging answers the following question: which fault candidate at which time step can be activated to cause the erroneous behavior of the corresponding erroneous trace?

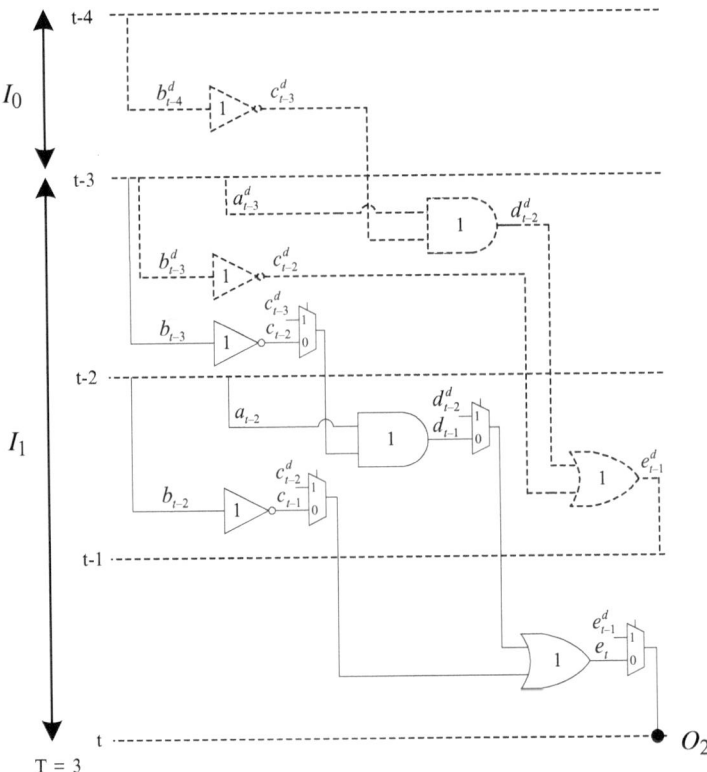

Fig. 7.6 TAM_DBG instance when T = 3 and D = 1

7.4.1 Algorithm

The debugging algorithm is shown in Fig. 7.7 in pseudocode. A TAM circuit, an erroneous trace ET, and a maximum variation delay D are the inputs of the algorithm. The output of the debugging algorithm is a set of fault candidates $\mathscr{F}$. Each fault candidate $F_i \in \mathscr{F}$, $i = 1, 2, \ldots, |\mathscr{F}|$, contains the location and the time step of a gate.

The TAM circuit is copied D times in line 2. Then, multiplexers are inserted to model timing variations (line 3). The set SEL denotes the set of select lines of multiplexers. A variable $sel_i \in SEL$, $i = 1, 2, \ldots, m$, represents the integer value of the select lines related to multiplexer i. Line 5 constrains the CNF by parameters of the erroneous trace ET. Variable d is the variable for limiting the currently considered timing variation which is initialized to 1 (line 7). The constraint of line 10 guides debugging to find timing variations explaining the erroneous behavior under a timing variation of d. If the CNF is satisfiable, all solutions are extracted and debugging terminates (lines 14–15). Otherwise, the previous constraint is removed

Fig. 7.7 TAM-based
debugging

```
1   algorithm  TAM_DBG (In : TAM, ET, D,  Out : 𝓕)
2   Copy (TAM, D)
3   SEL = Insert_Multiplexers()
4   // sel_i ∈ SEL, i = 1, 2, ..., m
5   Add_Constraint (ET : I_0, I_1, O_2, T )
6   𝓕 = ∅
7   d = 1
8   do
9   {
10      Add_Constraint ( (∑_{i=1}^{m} sel_i) = d )
11
12      if  Solve() == SAT  then
13      {
14          𝓕 = Extract_All_Solutions()
15          break
16      }
17      else
18      {
19          Remove_Constraint ( (∑_{i=1}^{m} sel_i) = d )
20          d = d + 1
21      }
22   } while  d ≤ D
23   end algorithm
```

(line 19). The limit d increases (line 20) in order to find a solution with an increased timing variation in the next step. This procedure repeats until d reaches the maximum timing variation D (line 22).

In the case of multiple erroneous traces, one TAM_DBG instance for each erroneous trace is created. To model transient faults, select lines of a single gate in different instances and in different time steps should be independent, because in each erroneous trace, some independent transient faults may be activated. Select lines of a single gate in different instances and in different time steps are connected to each other for permanent faults. A transient fault and a permanent fault can be distinguished by reapplying test vectors to recognize whether the erroneous behavior is observed again.

7.5 Experimental Results

TAM-based debugging is used in this section to experimentally debug logic circuits under timing variations. The experiments are carried out on a Quad-Core AMD Phenom(tm) II X4 965 Processor (3.4 GHz, 8 GB main memory) running Linux. The combinational and sequential circuits of ISCAS'85 and ISCAS'89 benchmark suites are used to evaluate our approach. The circuits are synthesized by Synopsys Design Compiler with Nangate 45 nm Open Cell Library [Nan11]. The TAM-based debugging described in this chapter is implemented using C++ in the WoLFram environment [SKF+09]. For the experiments, one time unit is 0.01 ns. We use MiniSAT as underlying SAT solver [ES04].

A simulation testbench is utilized to obtain the effect of timing variations on the outputs. The simulation testbench is implemented using Verilog in the ModelSim environment. In the simulation testbench, there are two instances of a circuit: a golden instance and a faulty instance. The outputs of these two instances are compared to detect an error and constitute an erroneous trace. A single slowdown fault of one time unit is injected in the circuit to create a faulty instance. Several points in the circuit are chosen as fault locations. At a clock period T, our approach generates random test vectors and applies them to the golden instance and the faulty instance of the circuit. If no error is observed for the activated faults, the clock period is decreased (clock shrinking). Having a new clock period, the procedure repeats until an erroneous behavior is observed. An erroneous trace is constituted by test vectors activating the fault at the specified frequency and the corresponding erroneous output values. The erroneous trace is given to TAM-based debugging. Having the initial erroneous trace, TAM-based debugging starts to find potential fault candidates.

The experimental results are presented in Table 7.1. The table shows the circuit name (first column), the total number of gates (#Gates), and the final number of fault candidates (#FC). Table 7.2 shows the required run time (Time) measured in CPU seconds (s).

In Section #FC of Table 7.1, column *Gate* indicates the total number of gates returned as fault candidates. In this column, each fault candidate is a gate indicating if a slowdown of one time unit at the appropriate time step on the output of the corresponding gate occurs, the erroneous behavior of the erroneous trace is created. Column *Path* counts the total number of possible paths constituted by the gates in column *Gate*. A path does not necessarily start at a primary input node reaching a primary output node. A path can be a segment in the middle of the circuit. The length of the shortest path from a real fault location to the fault candidates is shown in column *Dist*. Therefore, if the fault candidates contain the real fault location, the distance is zero. For single faults this must be the case, as the fault site would not be activated otherwise and, consequently, no erroneous behavior could be observed. A higher diagnosis accuracy is indicated by a smaller number of paths in *Path* with a smaller number of gates on them (*Gate*) with a shorter distance from a real fault location. The diagnosis accuracy can be increased by having higher quality erroneous traces, e.g., using a similar approach as the one suggested in Chap. 3 [DSF11]. One advantage of our approach is automatically extracting a segment of a path as fault candidate.

The total number of gates in the original circuit is shown by the second column in Table 7.1. By inserting buffers at the output of the original gates, the original circuit is converted to an untimed circuit. The total number of gates in the untimed circuits is shown in the third column. Afterwards, the TAM circuit is constructed. The fourth column shows the total number of gates in the TAM circuits. The column *UpperBound* in the table shows the upper bound to the size of the TAM as explained

Table 7.1 TAM-based debugging (size and diagnosis accuracy)

Circuit	#Gates when Time Unit = 0.01ns				#FC		
Comb.	Original	Untimed	TAM	Upper Bound	Gate	Path	Dist.
c17	6	26	35	35	2	1	0
c432	115	511	15446	151026	20	1	0
c499	179	840	4358	30476	2	1	0
c880	172	814	6483	13713	17	1	0
c1355	238	1112	14338	261280	26	2	0
c1908	142	658	5171	19896	3	1	0
c2670	280	1296	8817	16358	19	1	0
c3540	391	1792	50664	849976	6	2	0
c5315	632	3042	18283	49907	6	3	0
c7552	772	3657	58468	701207	22	1	0
Seq.							
s27	9	40	79	87	3	1	0
s298	59	277	893	1079	6	1	0
s386	67	300	575	1004	5	2	0
s444	83	370	1158	1450	7	1	0
s526	97	442	1235	1704	7	1	0
s713	96	421	3538	5063	13	1	0
s838	130	642	2250	2642	11	1	0
s953	216	967	3567	6403	9	1	0
s1196	280	1279	9124	18788	21	3	0
s1238	278	1278	10842	22581	15	1	0
s1494	315	1443	4373	6556	3	1	0
s5378	635	2884	8355	18505	7	1	0
s9234	813	3836	17473	42198	9	2	0
s15850	1537	7389	64402	2076798	5	1	0
s35932	3630	17728	18869	47393	4	1	0
s38584	6438	29651	89018	257023	1	1	0
Average	677.31	3180.58	16069.77	177044.15	9.6	1.3	0.0

in Sect. 6.3. In our experiments, the circuit c6288 of ISCAS'85 is omitted. This circuit is a multiplier known to have a very large number of paths and requires special approaches like [QW03] to handle the paths. The worst-case size of the TAM for the circuit $c6288$ is 2.16E+016. In this case, the process to build the TAM reaches the time limit of 48 h.

For circuit c17, the number of original gates is 6, while the number of untimed gates is 26. It shows that 20 (26–6) additional buffers have been added to the original circuit to create the untimed circuit.

The number of untimed gates for c499 is larger than for c432. But the number of TAM gates for c432 is larger than for c499. This shows that in c432 there are more paths in comparison to c499 which consequently increase the size of the

Table 7.2 TAM-based
debugging (time)

Circuit	Time (s)		
Comb.	TAM	DBG	Total
c17	0.00	260.63	260.63
c432	241.11	2156.11	2397.22
c499	13.58	294.75	308.33
c880	45.39	1796.25	1841.64
c1355	170.95	2889.47	3060.42
c1908	18.19	391.61	409.80
c2670	109.40	2022.07	2131.47
c3540	3063.39	1061.53	4124.92
c5315	504.08	827.67	1331.75
c7552	4008.44	3568.92	7577.36
Seq.			
s27	0.01	361.54	361.55
s298	0.20	647.91	648.11
s386	0.09	543.74	543.83
s444	0.53	749.76	750.29
s526	0.60	745.57	746.17
s713	18.55	1439.69	1458.24
s838	6.77	1167.48	1174.25
s953	16.85	990.54	1007.39
s1196	111.89	2370.67	2482.56
s1238	158.46	1729.87	1888.33
s1494	22.38	415.31	437.69
s5378	143.61	844.62	988.23
s9234	428.18	1159.88	1588.06
s15850	5478.62	1145.70	6624.32
s35932	1374.00	745.58	2119.58
s38584	11123.10	1198.25	12321.40
Average	1040.71	1212.50	2253.21

TAM. The number of fault candidates for c432 is 20 (20 gates), while the number of fault candidates for c499 is 2 (2 gates). In our experiments, all fault candidates together highlight failing speedpaths or some segments of failing speedpaths. Each fault candidate includes the location and the time step of fault activation.

Also the number of untimed gates for c5315 is larger than for c3540 while the number of TAM gates for c3540 is larger than for c5315. This indicates a larger number of paths in c3540. Therefore, the required time to create the TAM increases for c3540 which is shown in the table. In the table, usually debugging needs a longer time than the TAM creation process. But when a circuit has a large number of paths (for example c3540 and c7552), the time for TAM creation increases.

Our assumption is that all registers are observable for sequential circuits. In the table, #Gates for sequential circuits indicates the number of combinational

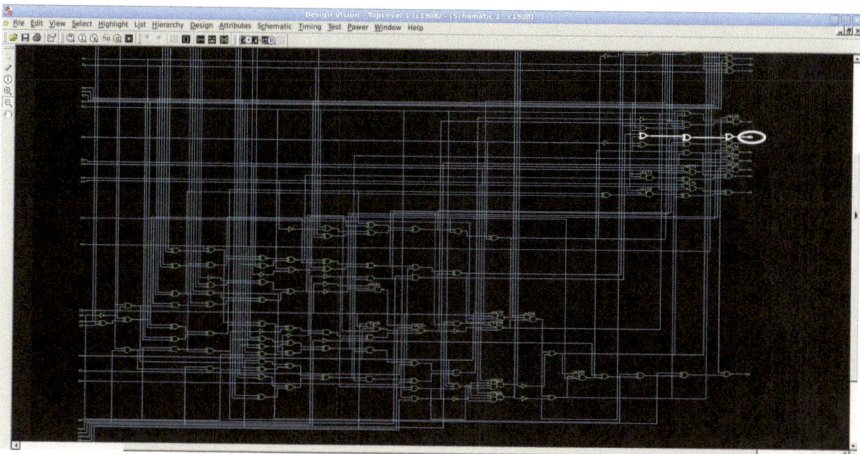

Fig. 7.8 Failing speedpath in circuit c1908

components excluding the number of flipflops. As the table shows, the debugging time for the most of sequential circuits is longer than the TAM creation time. For circuits s15850, s35932, and s38584 which are the larger circuits and have more paths, the TAM construction time is longer than the debugging time.

The average value for each column is calculated in the final row in the table. On average, the diagnosis accuracy of 98.59 % is achieved by our approach.

For circuit c1908, there are 3 fault candidates which are visualized on the schematic view of the circuit in Fig. 7.8 using *Synopsys Design Vision* [Syn14]. The gates highlighted in white color show a segment of a speedpath which has violated the timing constraint. The white oval shows the output on which the error was observed. For sequential circuits, the erroneous behavior may be observed on internal registers of the circuit. Figure 7.9 shows the failing speedpath for the sequential circuit s298. The white oval shows a flipflop on which the error was observed. As the experiments show, our approach achieves a high diagnosis accuracy and can automatically extract failing speedpaths.

7.6 Summary

This chapter introduced an approach to automate debugging for logic circuits under timing variations. In the framework, first the timing behavior of a circuit is converted into the functional domain under a discrete model of time unit. The new circuit is called TAM. Then, variation models are inserted in the TAM. Afterwards, TAM-based debugging finds potential fault candidates including their spatial and temporal information. The experimental results showed that the approach achieves 98.51 % diagnosis accuracy.

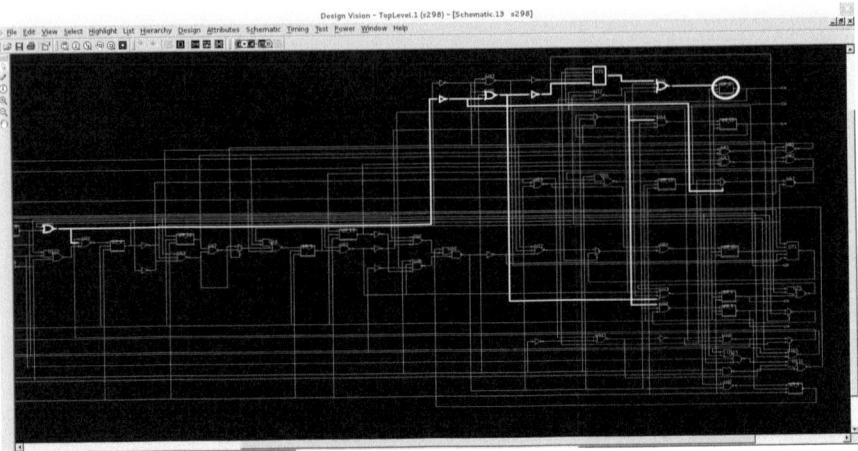

Fig. 7.9 Failing speedpath in circuit s298

All fault candidates highlight the sensitized paths leading to the error. The sensitized paths as failing speedpaths are visualized on the schematic view of the circuit. The debug team can focus only on the visualized fault candidates to find the root cause of the error.

In the next chapter, we investigate how to reduce the size of the debugging instance in order to decrease the time of debugging.

Chapter 8
Efficient Automated Speedpath Debugging

There are some approaches which try to reduce the size of the debugging model in order to efficiently localize the root causes of an error. QBF is used in [ASV+05] to reduce the size of the debugging instance. Moreover, the performance and applicability of debugging is improved using MaxSAT which simplifies the formulation of the debugging problem and reduces the size of the debugging instance and the debug time [CSMSV10]. Abstraction and refinement techniques can also handle large designs with a better performance by debugging an abstract model of the circuit [SV07]. The X value (three-valued logic) is used to abstract a circuit for efficient model checking [GSY07]. However, the previous approaches do not consider any timing information of the circuit.

TAM-based debugging presented in the previous chapter uses copies of a gate to represent the value of a gate at different points in time. Using multiple copies of a gate increases the size of the model and consecutively increases the debugging time. In this chapter, we present an efficient approach to automate speedpath debugging which integrates *Static Timing Analysis* (STA) and functional analysis in order to efficiently construct a compact timing model of a circuit [DF13b]. As a result, the debugging time decreases significantly. The approach is called STA-based debugging. Given erroneous behavior observed on the circuit outputs due to timing variations, a debugging instance based on SAT is created to automatically extract potential failing speedpaths. The approach is also utilized to debug an overclocked circuit which may fail to produce the expected outputs under timing variations. In comparison to the work presented in Chap. 7, the experimental results on ISCAS'85 and ISCAS'89 benchmark suites show a 63 % decrease in the size of the model and 54 % decrease in the debugging time. At the same time, this new approach achieves the same diagnosis accuracy as the approach of Chap. 7.

The remainder of this chapter is organized as follows. Our approach is presented in Sect. 8.1. Section 8.2 explains effects of timing variations and overclocking on

© Springer International Publishing Switzerland 2015   115
M. Dehbashi, G. Fey, *Debug Automation from Pre-Silicon to Post-Silicon*,
DOI 10.1007/978-3-319-09309-3_8

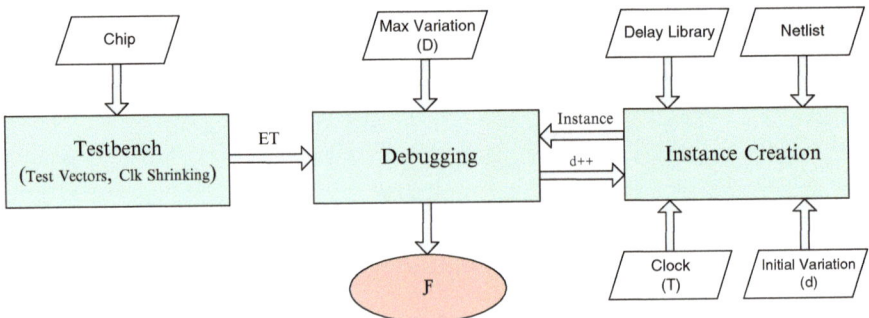

Fig. 8.1 Overview of proposed approach

the function of a circuit. Then in this section, the debugging model and algorithm
are demonstrated. Section 8.3 presents experimental results on benchmark circuits.
Section 8.4 summarizes this chapter.

8.1   Approach

Figure 8.1 shows an overview of our approach. Post-silicon validation of the
correct timing behavior involves applying test vectors to a chip and clock shrinking
[OHN09, KKC07]. This step is performed in a *testbench* in Fig. 8.1. When an error
is observed on the outputs or the flipflops, this error is returned as an *erroneous trace*
(ET). As mentioned in Sect. 7.1, an ET consists of the activating test vectors at the
specified frequency and the observed error and is denoted as $ET(I_0, I_1, O_2, T)$.

A slowdown increases the delay of some paths. These paths may violate the
frequency constraint and may create an erroneous trace. In this case, the goal is
finding speedpaths which have failed and have created the erroneous output of the
corresponding ET. To automate debugging, also the timing behavior of the circuit
in the presence of variations is converted into the functional domain. In Fig. 8.1,
the *Instance Creation* engine creates a functional model of timing behavior of the
circuit. The inputs of the instance creation engine are a netlist, a delay library, the
clock period T and the timing variation d. We discuss the instance creation engine
in Sects. 8.2.2 and 8.2.3 in detail.

Having the debugging instance and an ET, debugging begins. First the inputs
and output of the debugging instance are constrained according to the values of
the corresponding ET. Then debugging investigates whether a timing variation on a
gate is observable as the erroneous output value of the corresponding ET. If some
solutions are found by debugging, they are returned as fault candidates $\mathscr{F}$ and the
algorithm terminates. Otherwise, if no solution is found, the timing variation d
increases. Then a new instance according to the new timing variation is created. This
procedure repeats until the timing variation reaches maximum timing variation D.

In the following, first the effects of timing variations and overclocking on the function of a circuit are investigated (Sect. 8.2.1). Then the debugging model and algorithm are presented which are able to diagnose failing speedpaths under timing variations and overclocking.

8.2 STA-Based Debugging

8.2.1 Overclocking Versus Timing Variation

In this section, first the effect of overclocking on the function of a circuit is explained [VARR11]. Then, the effect of timing variations on the function of a circuit is investigated and compared to overclocking. Finally, a model is constructed which is able to convert the timing behavior of not only a normal circuit but also an overclocked circuit into the functional domain under timing variations.

Let D_l and T be the delay of the circuit and the clock period, respectively. Then the output at clock cycle $i + 1$ (O_{i+1}) depends on the inputs of the following clock cycles [VARR11]:

$$I_i, \quad I_{i-1}, \quad \ldots, \quad I_{i-b} \qquad b = \lceil D_l/T \rceil - 1$$

The clock cycles are indicated by the indices. For example, when $D_l = 10$ and $T = 9$, the output at clock cycle 2 (O_2) depends on the inputs of clock cycles 1 and 0 (I_1, I_0). In the remainder of the chapter, for the sake of simplicity, we discuss the case in which output O_2 depends on the inputs of two previous clock cycles I_1 and I_0. In this case, the signals on longer paths fail to arrive at outputs as their delay is longer than the clock period. Therefore, the output at clock cycle 2 (O_2) is affected by the inputs of clock cycle 1 (I_1) through shorter paths and the inputs of clock cycle 0 (I_0) through longer paths. The function of the circuit changes such that the output depends on a special combination of I_1 and I_0. In the following, it is explained how to construct the new function with an example.

In the original circuit of Fig. 8.2, first AT_{max} and AT_{min} are calculated by traversing the circuit from inputs to outputs (Sect. 6.3). In the figure, the arrival times are shown as a pair (AT_{min}, AT_{max}). Afterwards, by traversing the circuit from output to inputs through a path, the propagation time PT of the output of each gate along the path is calculated (Sect. 6.3). The output of a gate in the circuit is called a *point* in the circuit. Also, a PI or a PO is considered as a point in the circuit. The PT of a certain point along a particular path is the time required for a signal at that point to propagate to the PO. The propagation time PT of each point of the circuit is written after arrival times in Fig. 8.2. The number on each point p in the

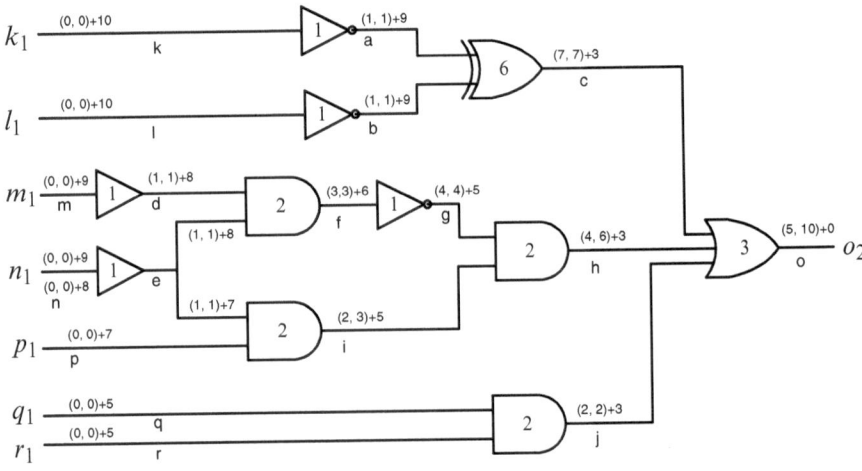

Fig. 8.2 Original circuit, T = 10

circuit denotes $(AT_{min}, AT_{max}) + PT_p^P$, where P is the corresponding path. During the backward traversal along a path P, three classes N, C and I may occur:

$$Class(p) = \begin{cases} N & \text{if } AT_{max} + PT_p^P \leq T, \\ C & \text{if } T < AT_{min} + PT_p^P, \\ I & \text{if } AT_{min} + PT_p^P \leq T < AT_{max} + PT_p^P. \end{cases} \qquad (8.1)$$

Classes N, C and I are also called *normal classes*. In contrast to normal classes, there are variation classes which consider the effect of timing variations and are discussed in the next section. Having the class of each point of the circuit, the approach creates the compact timing model of a circuit in the functional domain. In class N (*Non-critical class*), the signal at the point p has enough time to propagate to the PO along the corresponding path. Because the sum of the maximum arrival time from inputs to that point (AT_{max}) and the required time to propagate a signal from that point to the PO is smaller than clock period T. This point of the circuit is called non-critical point and is denoted by N. In this case, the function of this point of the circuit depends on only some inputs of I_1 which fall in the input cone of the corresponding point.

A critical point is implied by class C (*Critical class*). As in this point, no signal has enough time to propagate to the output, the function of this point depends on only some inputs of I_0 which fall in the input cone of the corresponding point. A combination of inputs I_1 and I_0 influences the function of signal s in class I (*Indefinite class*). The backward traversal starting at an indefinite point (I) always leads to independent C and N points.

Example 8.1. Considering the circuit of Fig. 8.3, when $T = 9$, the clock is overscaled because the clock period ($T = 9$) is less than the delay of the longest

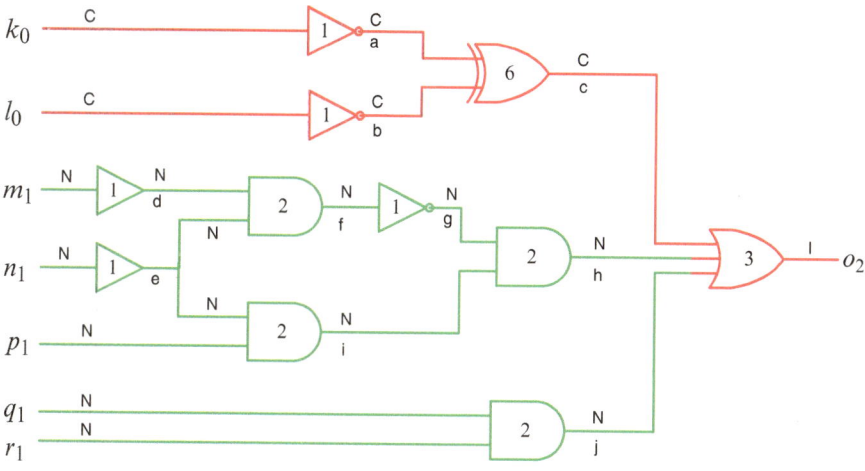

Fig. 8.3 Overclocked circuit, T = 9

path ($D_l = 10$). Therefore, output O_2 depends on a combination of I_1 and I_0. In this example, point o_2 satisfies the condition of class I and is marked as I. In Fig. 8.3, a point class is written on top of the wire. A lower-case letter under the wire indicates a wire name. By backward traversal of point o_2, this class is decomposed into classes C and N. Inputs k and l are critical and are considered at clock cycle 0. Other inputs are non-critical and considered at clock cycle 1. In the figure, non-critical parts are shown by green color. Other parts have red color.

At every point in the circuit, a timing variation may occur. In the example circuit of Fig. 8.4, the assumption is that there is a slowdown of one time unit at the output of gate a. Therefore, the delay of gate a increases from 1 to 2. When $T = 10$ and there is no timing variation, the circuit has normal behavior. But having a timing variation at point a, some paths of the circuit become critical. In this example, path (k, a, c, o) becomes critical. Therefore, input k is taken at clock cycle 0. Although the overclocking has the same effect on all paths, i.e., decrease of the clock period for all paths of the circuit, timing variation may affect only some special paths of the circuit, i.e., increase of the delay for some paths of the circuit. This is seen by comparing Figs. 8.3 and 8.4.

An overclocked circuit has some critical paths and some non-critical paths. In this case, a slowdown fault may change the function of a circuit more or less severely. Figure 8.5 depicts this case. In the example, T is 9 and there is a slowdown fault at the output of gate f. Therefore, more paths become critical. *If a section of the circuit has contributed in both a critical path and a non-critical path, that section of the circuit is duplicated.* In the example of Fig. 8.5, gate e contributes in one critical path (n, e, f, g, h, o) and one non-critical path (n, e, i, h, o). Thus, it is duplicated such that in the critical path the input is taken from clock cycle 0 (n_0) and in the non-critical path the input is taken from clock cycle 1 (n_1).

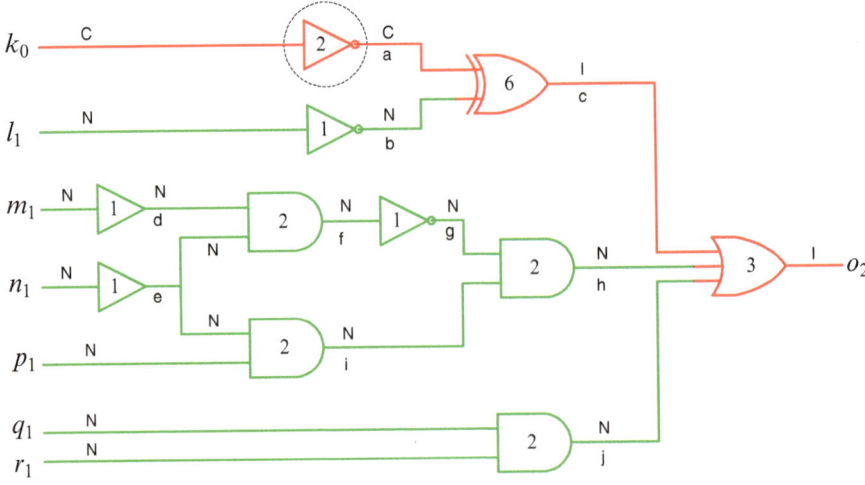

Fig. 8.4 Faulty circuit with a slowdown at gate a, T = 10

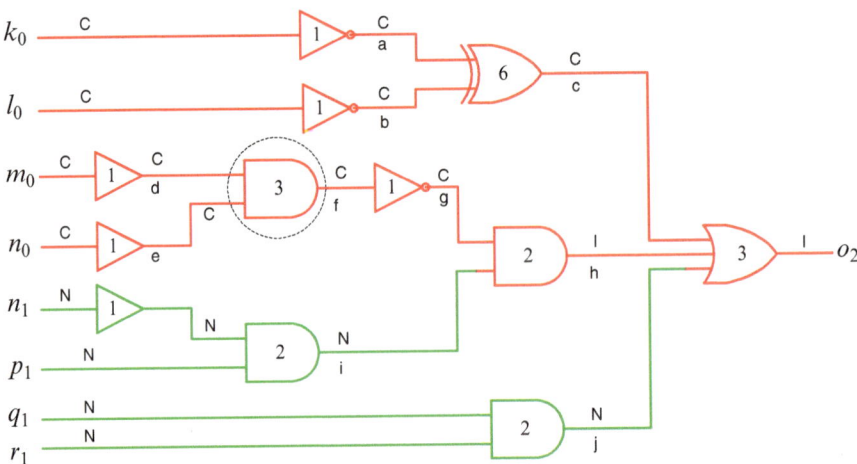

Fig. 8.5 Overclocked circuit with a slowdown at gate f, T = 9

Our assumption is that the function of a gate itself does not change. However, there can be more precise models which can also change the function of a gate. Assume that there is a signal which has been propagated and arrived at the middle of a gate and does not have enough time to propagate to the output of the gate. Therefore, the function of a gate can change depending on internal structure of a gate.

8.2.2 Debugging Model

An instance including timing variation models is called a debugging model. The model is able to activate a timing variation at every point of a circuit, i.e., it is able to change the function of each point of a circuit due to an activated timing variation.

We investigated the effect of a slowdown fault at a special point of the circuit in the previous section. However, a slowdown fault may occur at every point of a circuit and may change the function of the corresponding points and circuit outputs. Therefore, we add timing variation as a parameter to Formula (8.1) in order to obtain the timing behavior of each point of the circuit. The new formula is written as follows:

$$Class(p, d) = \begin{cases} N' & \text{if} \quad AT_{max} + PT_p^P + d \leq T, \\ C' & \text{if} \quad T < AT_{min} + PT_p^P + d, \\ I' & \text{if} \quad AT_{min} + PT_p^P + d \leq T < AT_{max} + PT_p^P + d. \end{cases}$$

$$(8.2)$$

The considered slowdown at point p is denoted by parameter d in this formula. Classes N', C' and I' are also called *variation classes*. These classes indicate the class of a point of a circuit when a timing variation (slowdown) occurs. In class N', even when having slowdown d, point p is a non-critical point. Class C' denotes a critical point when considering slowdown d. In class I', the function of point p, with considering slowdown d, depends on a combination of the inputs of clock cycle 1 and clock cycle 0.

The normal class of point p in a circuit can be N, I or C. While the variation class of point p can be N', I' or C', which show the class of point p when a timing variation occurs. Figure 8.6a shows the possible transitions from a normal class to a variation class. In the following, a transition is denoted by a dot. For example, a transition from class N to class N' is denoted by $N.N'$. When the normal class of a point is N, a slowdown increases the propagation time of the corresponding point. Therefore, class N may be converted to classes N', I' or C', i.e., transitions $N.N'$, $N.I'$ or $N.C'$ may occur. Also increasing the propagation time in class I may

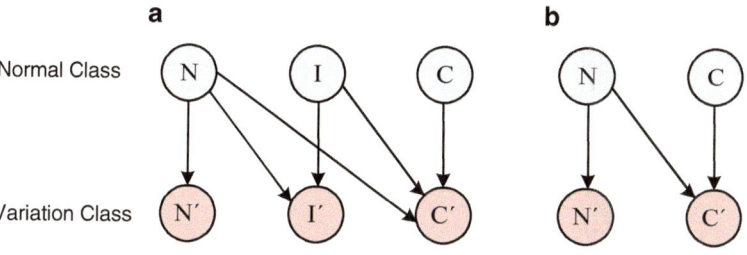

Fig. 8.6 Transitions from normal classes to variation classes for a point p: (**a**) p is an internal point, (**b**) p is a primary input

result in classes I' or C', i.e., transitions $I.I'$ or $I.C'$. The transitions including an indefinite normal class (I) or an indefinite variation class (I') are called *indefinite transitions*. $N.I'$, $I.I'$ and $I.C'$ are indefinite transitions. Other transitions are called *definite transitions*: $N.N'$, $N.C'$ and $C.C'$.

There is no indefinite class for a primary input because the arrival times are zero for primary inputs. Therefore, the primary inputs can have only non-critical and critical classes. Figure 8.6b shows the transition states for a primary input. *If a point of the circuit has the non-critical class (Class(p) = N), but it is converted into a critical point when considering timing variation d (Class(p, d) = C'), this variability is modeled by inserting a multiplexer at point p changing its behavior in order to debug the circuit.*

Example 8.2. In the circuit of Fig. 8.7, clock period T is 10. There can be a slowdown of one time unit in every point of the circuit. In this example, $Class(k) = Class(a) = N$, while $Class(k, 1) = Class(a, 1) = C'$. Therefore, a multiplexer at point k and one at point a is inserted in order to change the timing behavior of the corresponding points. Points l and b also have the same conditions. Moreover, these sections of the circuit are duplicated enabling to model the effect of both clock cycle 1 and clock cycle 0. Therefore, gates a and b are duplicated. The multiplexer at point a is activated if there is a timing variation at the corresponding point. In this case, the input is taken from clock cycle 0. Otherwise, the input is taken from clock cycle 1. From an input with a multiplexer on it, along the successor nodes,

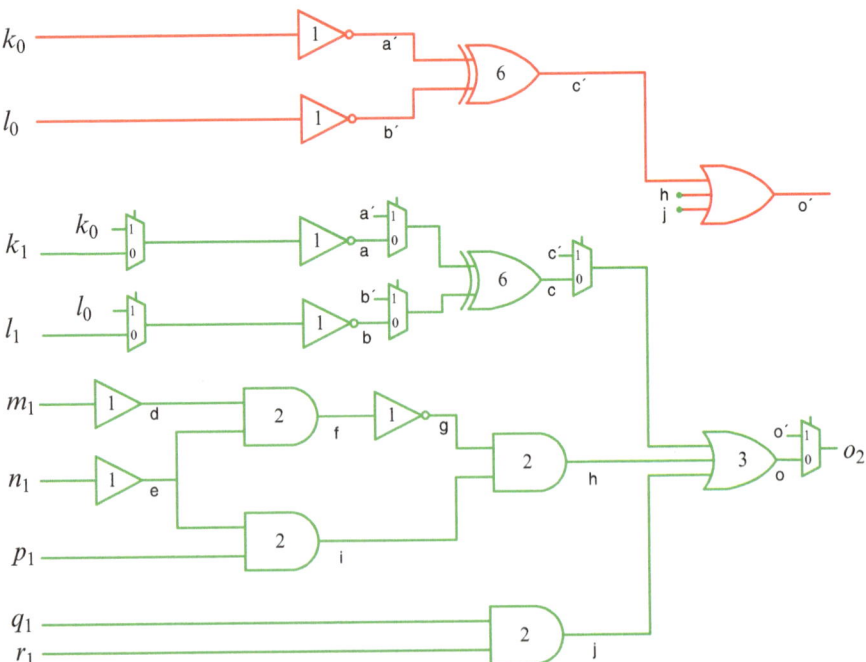

Fig. 8.7 Debugging instance, T = 10, slowdown = 1

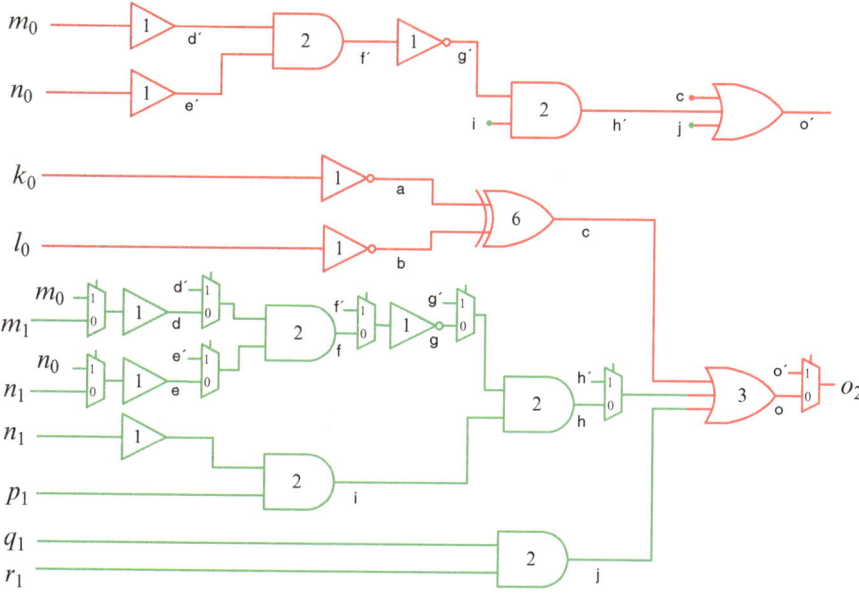

Fig. 8.8 Debugging instance, T = 9, slowdown = 1

the multiplexers are also inserted at the successor nodes whose normal classes or variation classes are indefinite, i.e., successor nodes with indefinite transitions.

As the debugging instance of Fig. 8.7 shows, a multiplexer modeling timing variation can be activated in order to change the function of the circuit according to a point at which a slowdown occurs. We call the paths including multiplexers *potentially-critical paths*.

Paths (k, a, c, o) and (l, b, c, o) are critical paths in the overclocked circuit of Fig. 8.8. In this case, a timing variation of one time unit renders some paths as potentially-critical paths. That means a timing variation on each point of the path can convert the non-critical path into a critical path. Paths (m, d, f, g, h, o) and (n, e, f, g, h, o) are potentially-critical paths. The gates on potentially-critical paths are duplicated in order to model the effect of timing variations in the functional domain.

The inputs and the output of the debugging instance are constrained to the inputs and the output values of the corresponding erroneous trace (ET) in order to debug the circuit. Then, the debugging engine answers the following question by activating the multiplexers: If there is a timing variation at a point of the circuit, can the erroneous behavior of the corresponding erroneous trace be observed? This investigation is performed by activating the select lines of multiplexers and observing its effect on the output. If the debugging instance is satisfiable, the debugging returns a set of fault candidates. Otherwise, timing variation d increases and a new instance is constructed until reaching maximum timing variation D.

8.2.3   Instance Creation Algorithm

The algorithm to create the debugging instance is shown in Fig. 8.9 in pseudocode. An original circuit, a delay library, clock period T and timing variation d are the inputs of the algorithm. In the algorithm the *original circuit* is traversed and a *new circuit* is constructed on the fly. A wire in the original circuit is called *original wire*. A wire in the new circuit is called *new wire*. For each wire of the original circuit, four flags are considered: $N.N'$, $C.C'$, $N.C'$, *Indefinite* (line 3). For example, if the flag $N.N'$ is 1, it shows that the original wire has already been visited through a non-critical path. In this case, the corresponding new wire in the new circuit has already been saved in the variable $w_{N.N'}$. For class $N.C'$, two variables are considered: one variable to save the new wire for the clock cycle 1 (non-critical wire $w_{N.C'}$), one variable to save the new wire for the clock cycle 0 (critical wire $w'_{N.C'}$).

The algorithm sets the flag $N.N'$ if an original wire is visited for the first time through a non-critical path. If this original wire is again visited through another non-critical path, the algorithm does not traverse the original wire again backward. Only its corresponding stored new wire is connected to the parent node. There is the same scenario for class $C.C'$ and class $N.C'$.

AT_{min} and AT_{max} are calculated by *Static Timing Analysis* (STA) in line 6. Then, the algorithm traverses the circuit backward starting at each *Primary Output* (PO) (lines 7–11). In line 10, the newly created outputs in the new circuit are stored in set *New_Out*. An original gate and its propagation time are the inputs of function *Traverse* (line 13). Function *Traverse* constructs a new gate (lines 36–42) after all of its inputs in the original circuit have already been visited. The output of function *Traverse* is the output of the newly created gate in the new circuit. In line 15, each input *in* of original gate *gate* is selected iteratively. The class of input *in* is calculated according to Formulas (8.1) and (8.2) (line 17). The classes are categorized into definite classes and indefinite classes. In the definite classes, the behavior of the point in the normal class and in the class of a timing variation is certain and fixed. Lines 19, 23, and 27 handle the definitive class. In the indefinite classes (line 31), the behavior of the point in the normal class or in the class of a timing variation is not known in advance, i.e., the function of this point of the circuit depends on a special combination of the inputs of clock cycle 1 and clock cycle 0. This combination may change depending on the propagation time of the current point in a path.

If a definite point has already been visited (lines 20, 24, and 28), it is not traversed again backward through a path with the same class. In this case, the algorithm uses only the stored wire of the corresponding point to create the parent node. Otherwise, the backward traversal continues (lines 21, 25, and 29). In the algorithm, whenever an indefinite class is visited, the backward traversal continues (line 33) until reaching a definite class.

A new gate is created after traversing all of the inputs of an original gate. If at least one of the gate inputs is a *Primary Input* (PI) with class $N.C'$ (potentially-critical point) (line 36), or one of the gate inputs is reached along a potentially-critical path (line 37), then two gates and one multiplexer are created in the new

```
 1   algorithm  DBG_Instance (In : Circ, DelayLib, T, d,  Out : Debug Instance)
 2
 3   foreach  Orig_Wire  create  Flags:  N.N', C.C', N.C', Indefinite
 4   foreach  Orig_Wire  create  New_Wires:  w_{N.N'}, w_{C.C'}, w_{N.C'}, w'_{N.C'}
 5
 6   STA ()
 7   foreach  out ∈ PO  do
 8   {
 9      gate = pre(out)
10      New_Out = New_Out ∪ Traverse(gate, 0)
11   }
12
13   function  W_out Traverse(gate, PT)
14   {
15     foreach  in ∈ gate.inputs  do
16     {
17       switch  Class(in).Class(in, d)
18       {
19         class  N.N':
20           if  N.N' == 1  then  W_in = W_in ∪ w_{N.N'}
21           else  W_in = W_in ∪ Traverse(pre(in), PT + gate.D)
22
23         class  C.C':
24           if  C.C' == 1  then  W_in = W_in ∪ w_{C.C'}
25           else  W_in = W_in ∪ Traverse(pre(in), PT + gate.D)
26
27         class  N.C':
28           if  N.C' == 1  then  W_in = W_in ∪ w_{N.C'} ∪ w'_{N.C'}
29           else  W_in = W_in ∪ Traverse(pre(in), PT + gate.D)
30
31         class  I.I' || I.C' || N.I':
32           // Indefinite
33           W_in = W_in ∪ Traverse(pre(in), PT + gate.D)
34       }
35     }
36     if  (∃in: in ∈ PI && in_{current_class} == N.C')  ||
37         (∃w: w ∈ W_in && w is mux_out)  then
38         W_out = W_out ∪ Create_Gate(W_in)
39         W_out = W_out ∪ Create_Gate'(W_in)
40         W_out = W_out ∪ Create_Mux()
41     else
42         W_out = W_out ∪ Create_Gate(W_in)
43
44     Set_Flags_and_New_Wires(gate.output, W_out)
45     return  W_out
46   }
47
48   end algorithm
```

Fig. 8.9 Debugging instance creation

circuit (lines 38–40). This case can be seen in the example of Fig. 8.7 where gate a, gate a', and a multiplexer are created. Otherwise, the algorithm creates one new gate (line 42). After creating the new gate, the output of the new gate is saved in its corresponding variable on the original gate, and also its corresponding flag is set (line 44).

In sequential circuits, the error may be detected several clock cycles after fault activation. In this case, the sequential circuit is unrolled as many times as the number of clock cycles constituting the erroneous trace. In each unrolled clock cycle, a debugging instance is created where its output depends on the inputs and state bits of two clock cycles.

Table 8.1 Speedpath debugging (size and diagnosis accuracy)

Circuit	#Gates				#FC		
Comb.	Original	TAM	NEW	%Decrease	Gate	Path	Dist.
c17	6	16	9	43.75	2	1	0
c432	115	7128	3089	56.66	20	1	0
c499	179	1824	1429	21.66	2	1	0
c880	172	2860	534	81.33	17	1	0
c1355	238	6044	4215	30.26	26	2	0
c1908	142	2240	276	87.68	3	1	0
c2670	280	3980	776	80.50	19	1	0
c3540	391	22934	5314	76.83	6	2	0
c5315	632	7714	1381	82.10	6	3	0
c7552	772	24738	15533	37.21	22	1	0
Seq.							
s27	9	36	29	19.44	3	1	0
s298	59	414	125	69.81	6	1	0
s386	67	262	146	44.27	5	2	0
s444	83	554	184	66.79	7	1	0
s526	97	568	219	61.44	7	1	0
s713	96	1726	592	65.70	13	1	0
s838	130	956	340	64.44	11	1	0
s953	216	1670	367	78.02	9	1	0
s1196	280	4244	591	86.07	21	3	0
s1238	278	5046	742	85.30	15	1	0
s1494	315	2078	398	80.85	3	1	0
s5378	635	3800	1556	59.05	7	1	0
s9234	813	7894	3621	54.13	9	2	0
s15850	1537	26258	3230	87.70	5	1	0
s35932	3630	7606	4178	45.07	4	1	0
s38584	6438	40188	18214	54.68	1	1	0
Average	677.31	7029.92	2580.31	63.30	9.6	1.3	0.0

8.3 Experimental Results

Our proposed debugging approach is used to debug logic circuits under timing variations in this section. We use the same experimental setup as Sect. 7.5.

Table 8.1 presents the experimental results. The table shows the circuit name (first column), the total number of gates (#Gates), and the final number of fault candidates (#FC). The required run time (Time) measured in CPU seconds (s) is presented in Table 8.2. In the tables, a column with name *TAM* indicates the results of the approach presented in Chap. 7. A column with name *NEW* indicates the approach presented in this chapter. The best results in the tables are marked bold.

Table 8.2 Speedpath debugging (time)

Circuit	Instance Creation Time (s)		Debugging Time (s)		Total Time (s)		
Comb.	TAM	NEW	TAM	NEW	TAM	NEW	%Decrease
c17	0.00	0.00	260.63	267.22	**260.63**	267.22	−2.53
c432	241.11	**0.16**	2156.11	1999.59	2397.22	**1999.75**	16.58
c499	13.58	**0.10**	294.75	287.52	308.33	**287.62**	6.72
c880	45.39	**0.09**	1796.25	1631.90	1841.64	**1631.99**	11.38
c1355	170.95	**0.19**	2889.47	2462.80	3060.42	**2462.99**	19.52
c1908	18.19	**0.04**	391.61	372.61	409.80	**372.65**	9.07
c2670	109.40	**0.32**	2022.07	2044.68	2131.47	**2045.00**	4.06
c3540	3063.39	**0.40**	1061.53	674.73	4124.92	**675.13**	83.63
c5315	504.08	**0.89**	827.67	768.74	1331.75	**769.63**	42.21
c7552	4008.44	**4.26**	3568.92	2332.99	7577.36	**2337.25**	69.15
Seq.							
s27	0.01	**0.00**	361.54	374.11	**361.55**	374.11	−3.47
s298	0.20	**0.01**	647.91	652.75	**648.11**	652.76	−0.72
s386	0.09	**0.01**	543.74	546.38	**543.83**	546.39	−0.47
s444	0.53	**0.03**	749.76	738.20	750.29	**738.23**	1.61
s526	0.60	**0.02**	745.57	754.66	**746.17**	754.68	−1.14
s713	18.55	**0.06**	1439.69	1301.63	1458.24	**1301.69**	10.74
s838	6.77	**0.11**	1167.48	1123.31	1174.25	**1123.42**	4.33
s953	16.85	**0.11**	990.54	960.95	1007.39	**961.06**	4.60
s1196	111.89	**0.12**	2370.67	2078.69	2482.56	**2078.81**	16.26
s1238	158.46	**0.12**	1729.87	1528.15	1888.33	**1528.27**	19.07
s1494	22.38	**0.09**	415.31	374.21	437.69	**374.30**	14.48
s5378	143.61	**2.81**	844.62	752.11	988.23	**754.92**	23.61
s9234	428.18	**6.24**	1159.88	951.07	1588.06	**957.31**	39.72
s15850	5478.62	**41.89**	1145.70	580.37	6624.32	**622.26**	90.61
s35932	1374.00	**249.00**	745.58	567.67	2119.58	**816.67**	61.47
s38584	11123.10	**423.06**	1198.25	264.74	12321.40	**687.80**	94.42
Average	1040.71	**28.08**	1212.50	1015.07	2253.21	**1043.15**	53.70

The total number of gates in an original circuit, TAM and our instance is shown in columns 2–4 in Table 8.1. It should be mentioned that the number of TAM gates in this table is the number of gates without considering the buffers in the TAM. We ignore the number of buffers in the TAM as the buffers do not have significant effect on the time of SAT solving. Column 5 in Table 8.1 shows the amount of decrease in the size of the *NEW* model in comparison to the *TAM* model in percent. The times in Table 8.2 are indicated as the required time to create a debugging instance (*Instance Creation Time*), the required time to debug (*Debugging Time*) and the total time. The last column in Table 8.2 shows the amount of decrease in the total time of the *NEW* approach in comparison to the *TAM* approach in percent.

As seen in Section #*FC* of Table 8.1, the diagnosis accuracy of the *NEW* approach is same as the *TAM* approach for single faults. The number of gates in our model is always smaller than the number of gates used in the approach presented in the previous chapter (comparison of columns 3 and 4 in Table 8.1). Therefore, the memory consumption of our approach is less than the TAM approach. On average, our approach requires 2580 gates while the TAM requires 7030 gates which implies the new approach has 63 % decrease in the size of the model in comparison to the TAM.

The required time to create a debugging instance in the new approach is always shorter than the TAM approach. This decrease of the instance creation time is tangible especially for large and complex circuits. The debugging time depends on not only the size of the model but also on the number of fault candidates. When the number of fault candidates is larger, the solver (debugging engine) needs a longer time to extract fault candidates. The total time is the sum of the instance creation time and the debugging time. For the large circuits, the new approach needs a shorter total time as the table shows. On average, the new approach spends 1,043 s while the TAM approach spends 2,253 s which indicates a 54 % decrease in the required total time.

The number of gates as fault candidates is 17 for circuit c880. These fault candidates constitute one path. While for circuit c1355, there are 26 gates as fault candidates constituting 2 paths. Also the distance is zero indicating the set of fault candidates includes the real fault location. For all circuits, the diagnosis accuracy of the new approach is same as the TAM approach for single faults.

8.4   Summary

In this chapter, we introduced a new and efficient approach to automate debugging for logic circuits under timing variations. The approach was based on converting the timing behavior of a circuit and its corresponding timing variations into the functional domain. Our approach integrated static timing analysis and functional analysis in order to build a compact debug instance. Having the new debug instance in the functional domain and an erroneous trace, potential failing speedpaths are automatically found by our debugging approach. The experimental results on ISCAS'85 and ISCAS'89 benchmarks suites showed a 63 % decrease in the size of model resulting in 54 % decrease in the debugging time in comparison to the fully flattened TAM model. The diagnosis accuracy remains at the high quality achieved by the previous approach.

Part III
Debug of Transactions

Chapter 9
Online Debug for NoC-Based Multiprocessor SoCs

Modern high-performance SoCs contain many IP cores such as processors and memories. NoCs have been proposed as a scalable interconnect solution to integrate large multiprocessor SoCs [BM02, PGI+05]. Having a large SoC with complex communication among its cores, the complete verification coverage at pre-silicon stage is almost impossible. Therefore in addition to electrical bugs, some design bugs may also appear in the final prototype of an SoC.

There are some previous works which present infrastructures for SoC debug. The existing debug infrastructures for complex SoCs are reviewed in [HMM06]. These infrastructures support debugging such that the internal nodes become observable and controllable from outside. A DFD technique for NoC-based SoCs is presented in [YPK10]. The technique enables data transfer between a debugger and a *Core-Under-Debug* (CUD) through the available NoC to facilitate debugging. The work in [TX07] proposes a debug platform to support concurrent debug access to the CUDs and the NoC in a unified architecture. This platform is realized by introducing core-level debug probes in between the CUDs and their network interfaces and a system-level debug agent. A ring-based NoC architecture is proposed in [TGR+12] to debug SoCs. The NoC is used to send back the information observed by monitors to the debugger. A *Non-Uniform Debugging Architecture* (NUDA) is proposed in [WCCC12] to debug many-core systems. A NUDA node in each cluster has three main parts: nanoprocessor, memory and communication. The NUDAs are distributed across a set of hierarchical clusters and are connected to each other through a ring interconnection. Then the address space is monitored using non-uniform protocols for race detection. Monitoring the address space without abstraction consumes a large storage and increases the latency of the error detection.

The main focus in most of the previous works is NoC test and diagnosis. Packet address driven test configurations are utilized in [RGU09] to test and to diagnose regular mesh-like NoCs using a functional fault model. Then, link faults are diagnosed using test results and a diagnosis tree. Interconnect faults in Torus NoCs are detected and diagnosed using BIST structures in [CAK+09]. Afterwards,

© Springer International Publishing Switzerland 2015
M. Dehbashi, G. Fey, *Debug Automation from Pre-Silicon to Post-Silicon*,
DOI 10.1007/978-3-319-09309-3_9

the NoC is repaired by activating alternative paths for faulty links. An NoC with a faulty router or a broken link is repaired in [CCLL11] using spare routers. The inherent structural redundancy of the NoC architecture is exploited in a cooperative way to detect the faults using BIST [SRL+11]. A fault is also localized by utilizing diagnosis units in switches. In the diagnosis unit there are different comparators to compare data from all the possible pairs of switch input ports. A comprehensive defect diagnosis for NoCs is proposed in [GPS+12]. The approach utilizes an end-to-end error symptom collection mechanism [SGC11] to localize datapath faults and a distributed counting and timeout-based technique to localize faulty control components [GPS+12]. The work in [AKSN07] diagnoses the NoC switch faults using hardware redundancy in each switch and a high level fault model. These works focus only on electrical bugs in the components of an NoC. However, we consider design bugs which influence communications in an NoC-based SoC.

The idea of transaction-based communication-centric debug is introduced in [GVVSB07] to debug complex SoCs which interact through concurrent interconnects such as NoC. The transactions are observed using monitors [VG09] and the debug control unit can control the execution of the SoC (stopping, single stepping, etc). In [GF09], transactions are stored at run time in a trace buffer using on-chip circuits. After executing a run of the SoC, the content of the trace buffer is read and analyzed offline using software. The analysis software tries to find certain patterns [EEH+06] in the extracted transactions that are defined by their *Transaction Debug Pattern Specification Language* (TDPSL). Because of limited size of a trace buffer, getting an execution trace of the transactions related to the time of bug activation is a challenging problem. To overcome this problem, the content of the trace buffer is utilized to backtrace the transactions along their execution paths [GF12]. The backtracing is performed in transaction-level states using BMC. However, backtracing needs formal pre-image computations which can blow up for large and complex designs [dPNN+11]. To address this problem, we need to have online detection to stop the SoC close to the time of bug activation at the transaction level.

In this chapter, a transaction-based debug infrastructure is presented which can be used not only for online debug and online system recovery but also for interactive debug in which an external debug platform programs the FSMs and the filters according to the considered assertions at each round of debugging [DF14b]. Our hardware infrastructure contains monitors, filters, and a debug network including *Debug Units* (DU). Filters and DUs are programmed according to the transaction-based assertions defined by TDPSL. Master interconnects are monitored to extract transactions. Slaves send information to masters. This redundant information is used to observe the elements of transactions online. No modification of the internal components of the NoC is required. At run time, the programmable FSMs in the DUs investigate the assertions online and detect an error.

The main contributions of this chapter are as follows:

- Proposing a debugging infrastructure to transaction-based online debug of NoC-based SoCs without modifying the internal components of the corresponding NoC (non-intrusive to the NoC).
- Analyzing and finding transaction-based debug patterns at run time using debug units including programmable filters and FSMs.
- Presenting an ordering mechanism in the routers of the debug network to order the transactions online.
- Online system recovery without stopping and interrupting the NoC.

The efficiency of our approach is shown in the experimental results using different assertion patterns defined by TDPSL such as race, deadlock, and livelock. An NoC-based SoC using a mesh network is setup in the Nirgam NoC simulator [Jai07] to evaluate our approach. Also, the effectiveness of the proposed online recovery is shown in the experimental results.

The remainder of this chapter is organized as follows. Our debug approach including hardware and software parts is explained in Sect. 9.1. Section 9.2 proposes a debug flow integrating the debug approaches of Parts I and II. The debug patterns and their corresponding FSMs are explained in Sect. 9.3. This section also presents experimental results on an NoC-based SoC. The last section concludes the chapter.

9.1 Approach

Transaction-based online debug aims at improving the observability and the controllability of the system. Whenever transactions conform to certain debug patterns, an error is detected. In this case, the DU sends the debug packets to the SoC nodes in order to control the network and to recover from the error state.

The following requirements are defined to enable transaction-based online debug:

1. Debug infrastructure has to be able to collect the elements of each transaction at run time.
2. Transactions have to be ordered at run time.
3. Debug patterns, i.e., the relation of transactions, have to be asserted at run time.

If a debug infrastructure fulfills the three mentioned requirements, it can be used for transaction-based online debug.

In the following, we explain our debug infrastructure fulfilling the above mentioned requirements. To collect all elements of each transaction in a system based on NoC (first requirement), we require distributed monitors and *Debug Redundant Information* (DRI). Monitors and DRI are explained and discussed in Sects. 9.1.1 and 9.1.2, respectively.

The transaction ordering mechanism in the DU is responsible to order transactions (Sect. 9.1.4) fulfilling the second requirement. A DU is the main part of the debug infrastructure which searches for certain debug patterns in the received transactions. We utilize a tree-based debug network structure in which all monitors have a short distance to DUs. In the debug network, the transactions are ordered using DUs. The ordered transactions are transferred on each link of the debug network from bottom to top such that the ordered transactions can be utilized in each level of the debug network for hierarchical and assertion-based debug.

FSMs in DUs are utilized to investigate transaction-based assertions at run time to fulfill the third requirement (Sect. 9.1.6). The filters in DUs and in monitors help FSMs by dropping unrelated transactions (Sect. 9.1.5).

The hardware infrastructure of our approach for an SoC including four IPs, 2 masters and 2 slaves, is shown in Fig. 9.1. The debug infrastructure has the following parts: monitors, filters, DU, and DRI. The internal structure of a DU has also three main parts: transaction ordering, filter, and FSM. A *Network Interface* (NI) is used to connect each IP to the NoC in Fig. 9.1.

The tree-based debug infrastructure is shown in Fig. 9.2. The lowest level of the infrastructure includes monitors and filters. Monitors are connected to master interconnects (MI) to observe the transactions. The tree-structure is composed of two types of DUs which are one *Central Debug Unit* (CDU) and several *Local Debug Units* (LDU). A CDU is only at the top level. The other levels have LDU. LDU and CDU structures are explained in Sect. 9.1.3.

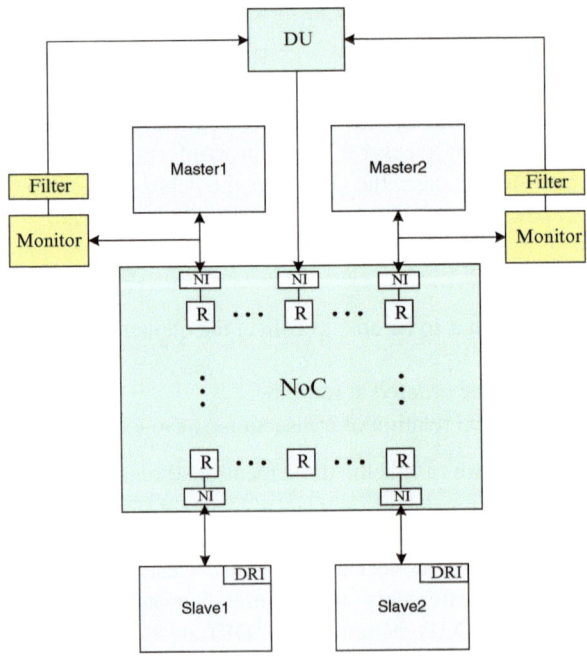

Fig. 9.1 Debug infrastructure

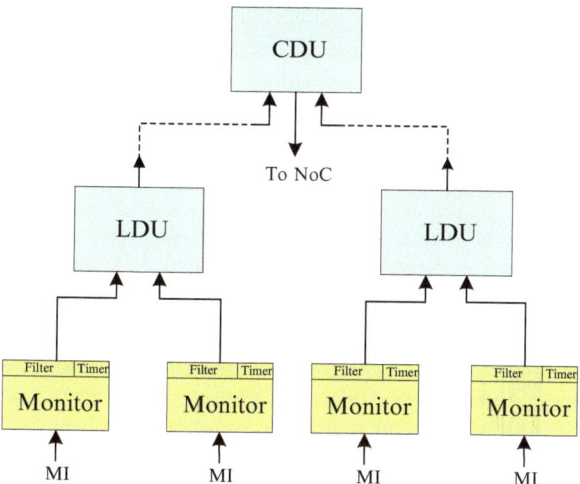

Fig. 9.2 Tree-based debug infrastructure

9.1.1 *Monitor*

The basic elements of a transaction are extracted by monitors as mentioned in Sect. 2.5.2. They observe master interconnects to enable transaction-based debug [GVVSB07]. In a packet-based protocol in an NoC, we can immediately extract the elements *master*, *slave*, and *type* by observing the master interconnects. But to extract the element *address* as SAME, SEQ, and OTHER, which is a comparison of the slave address in the current transaction and the previous transaction for the corresponding slave, we need some DRI. The next subsection explains the DRI.

A master interconnect is observed by a monitor in our infrastructure. The monitor signals a matching transaction expression explained in Sect. 2.5.2 as an output.

A timer is also included in each monitor. The timer is used to attach a timestamp to each observed transaction. The timestamp attached to a packet is utilized at DUs to order the transactions arriving from the left and right input links of DUs. As the transactions are consumed online using FSMs, large timestamps are not required. Timestamps only need to distinguish the order of transactions arriving at DUs.

In the case of an SoC with asynchronous IPs, the CDU sends synchronization packets to monitors. The timers in monitors are synchronized according to the synchronization packets. As only the CDU sends the packets from the top level to the bottom level in the tree-based debug network, the delay of synchronization packets arriving at monitors are predetermined. In this case, the time in monitors is synchronized with the time in the CDU by incrementing the CDU time included in the synchronization packet with the delay of the synchronization packet.

A monitor can also include some assertions to check the behavior of the corresponding IP. This kind of assertions is called *individual assertions* as they check only the behavior of an individual IP. The individual assertions do not only check the main elements of the transactions but also other data and events related to the corresponding IP. For example, the address field of a transaction can be checked to verify if the address is in an expected address range. Also some data dependencies can be checked by individual assertions for the corresponding IP.

9.1.2   Debug Redundant Information

The element *address* of a transaction is extracted and transferred by DRI. We can form the element *address* using slaves (slave-based approach) or using DUs (DU-based approach). In the following, these two approaches are discussed.

In a slave-based approach, the element *address* is formed in the slaves and is sent as redundant information to masters through the NoC. Because the element *address* is a comparison of the address of the current transaction with the address of the previous transaction for the corresponding slave, this comparison can be simply done in each slave.

Two bits redundant information are sent by the slave to the masters. These two bits specify the symbols SAME, SEQ, and OTHER. We can also use more symbols for the slave address to have more accurate data depending on the applications running on the SoC. We use the slave-based approach to detect deadlock and livelock. As in the deadlock/livelock assertions (Sect. 9.3.2.1) we need address symbols only for EoTrs, the salve-based approach is suitable. In the slave-based approach, address symbols are included in EoTrs.

The DRI section in each slave in Fig. 9.1 compares the slave address of the current transaction with the slave address of the previous transaction in the corresponding slave. Then the DRI section selects a symbol (SAME, SEQ, or OTHER) and adds this symbol in the response packet as two redundant bits. On the master side these two bits are read by monitors to constitute a complete transaction expression. These two redundant bits are used only in monitors. The master applications ignore these two redundant bits.

In this case, we can have the address information for the corresponding slave only in the *EoTr*. The element *address* is not available in *SoTr*. To have this information, we need to wait until receiving an *EoTr* by the corresponding slave.

The second approach to form the element *address* uses DUs, i.e., the DU-based approach. In this approach, monitors observe slave addresses and send them to the DUs. In the DU, there is one address register for each slave. The address registers keep the address of the previous transaction for each slave independently. When a new transaction is performed, the content of the address register related to the corresponding slave is compared to the new transaction address. Then the symbols SAME, SEQ, and OTHER are derived and the address register is updated to keep the slave address in the latest transaction for the corresponding slave. We use the DU-based approach to detect races.

When the DU-based approach is used, the element address is available for both *SoTr* and *EoTr*. The DU-based approach requires more memory in the DUs storing

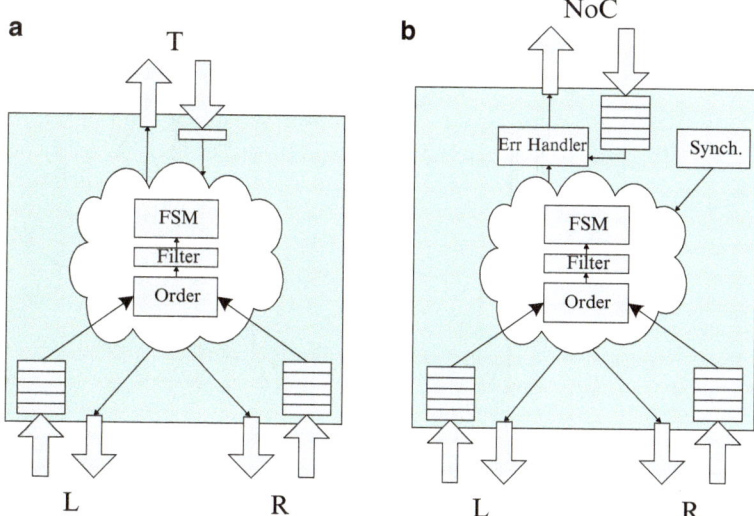

Fig. 9.3 (**a**) LDU structure, (**b**) CDU structure

slave addresses. Also it needs more bandwidth for the debug network to transfer slave addresses to DUs. The advantage of this approach is being non-intrusive to the SoC.

9.1.3 Debug Unit

A DU can be an LDU or a CDU. As shown in Fig. 9.2, a CDU is used at the top level in the tree-based debug network. An LDU is used in other levels of the debug network. The structure of an LDU is suitable to build a tree-based network. An LDU has three ports (Fig. 9.3a): top port T, left port L, and right port R. The right and left ports transfer the data observed by monitors towards the top level. Also the synchronization packets sent by the CDU are transferred from the top port to the left and right ports to finally reach the timers in the monitors. The CDU controls the traffic of the packets sent from the top level to the leaves in the debug network. For this data, we use only one buffer in each LDU to transfer synchronization packets.

The right and left FIFOs store the packets arriving at the inputs of the right and left ports. Then the transaction ordering selects a transaction packet such that the transactions are ordered based on their timestamps. The filter does not allow a transaction to be forwarded if the transaction is not related to the considered assertions. The related transactions regarding the considered assertions are used in the FSM to investigate the assertions. If an assertion fails, an error message is sent to the CDU.

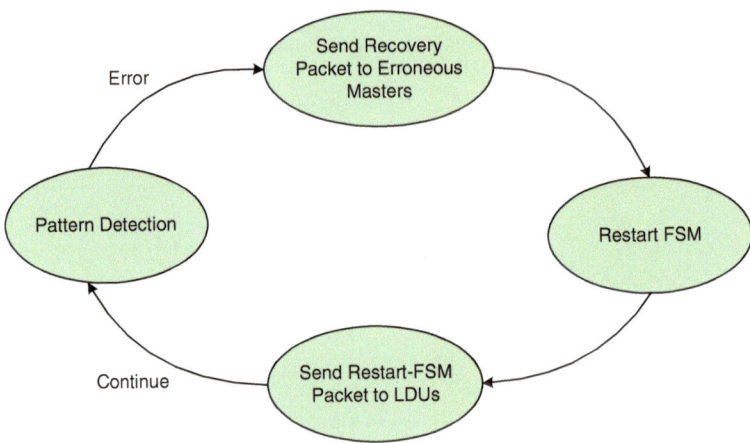

Fig. 9.4 CDU procedure to recover the SoC

The CDU is utilized at the top level of the debug network. The CDU has two additional tasks: synchronizing the timers and handling error cases (Fig. 9.3b). As Fig. 9.3b shows, synchronization is performed by part *Synch*. When there is an error, the error handler in the CDU manages the network by sending some debug packets to other nodes in the SoC. The CDU is connected to the NoC communicating with other nodes in the network. By this, the CDU can send the error state to all nodes or some special nodes in the network in order to have more accurate debug information or to recover the SoC from the error state.

The CDU procedure to recover the SoC from an error is demonstrated in Fig. 9.4. At the step of pattern detection, the CDU checks the debug patterns at run time. If the CDU detects an error, the second step starts. In this step, the CDU sends a recovery packet to the masters which have contributed to the observed error. A recovery packet contains an error type and additional information helping the masters to start a recovery process. At the third step, the CDU restarts its own FSM. Then, the CDU sends a restart packet to the LDUs restarting the LDU-FSMs. Afterwards, the procedure continues with the step of pattern detection.

Example 9.1. An example for a master recovery process in the case of a software deadlock is shown in Fig. 9.5. The process starts, when a master receives a recovery packet from the CDU with the error type deadlock. The master releases the locked resources in the first step. Then the master waits for a random time and proceeds its main function again. With this procedure, the system is online recovered from the error state without stopping and interrupting the NoC.

DUs include FSMs to investigate transaction-based assertions at run time. To check an assertion in an efficient way, both LDU-FSMs and CDU-FSM must be programmed. Distributed online assertion checking can be performed through programming the FSMs in different levels.

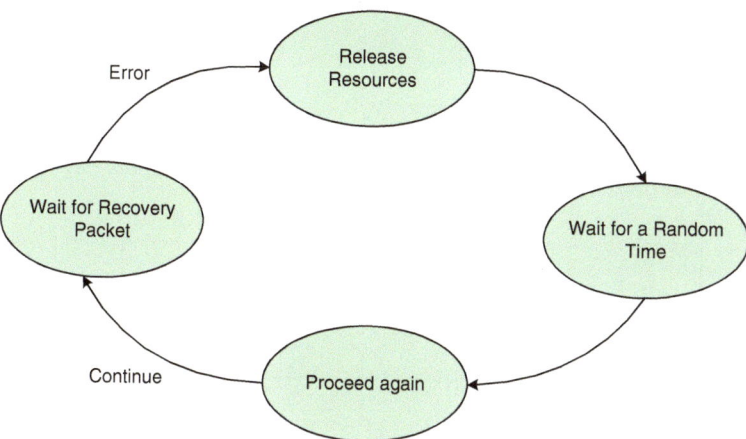

Fig. 9.5 An example for master recovery thread in the case of a software deadlock

In contrast to individual assertions, *distributed assertions* are used in the DU to check the distributed behavior of the corresponding SoC. The distributed assertions check the relation of transactions transferred between different IPs in the SoC.

9.1.4 Transaction Ordering

When there is more transaction traffic on one link than another link in the debug network, some early-generated transactions are accumulated and buffered in the FIFO of the corresponding DU. This case may also occur when the bandwidth of the debug network is less than the bandwidth of the NoC. When there are some transactions in the FIFO of the left link which have been generated earlier than the transaction available in the FIFO of the right link, the transaction of the right link has to wait until the transactions with smaller timestamps on the other link have been transferred. By comparing the timestamp of a packet in the left FIFO and the right FIFO, the packets are ordered based on their generation time.

The length of the timestamp depends on the worst case delay of the debug network. A timestamp needs only to distinguish the packets based on the time in which they have been generated or sampled. The part *Order* in DU in Fig. 9.3a,b compares the timestamps of a packet in the right and the left FIFOs. Then it selects a packet which has a smaller timestamp.

The accuracy of the debug pattern detection may be influenced by the size of the left and right FIFOs because if the FIFO becomes full, some transactions are lost. In this work, the assumption is that the size of the FIFOs is sufficient to process the transactions.

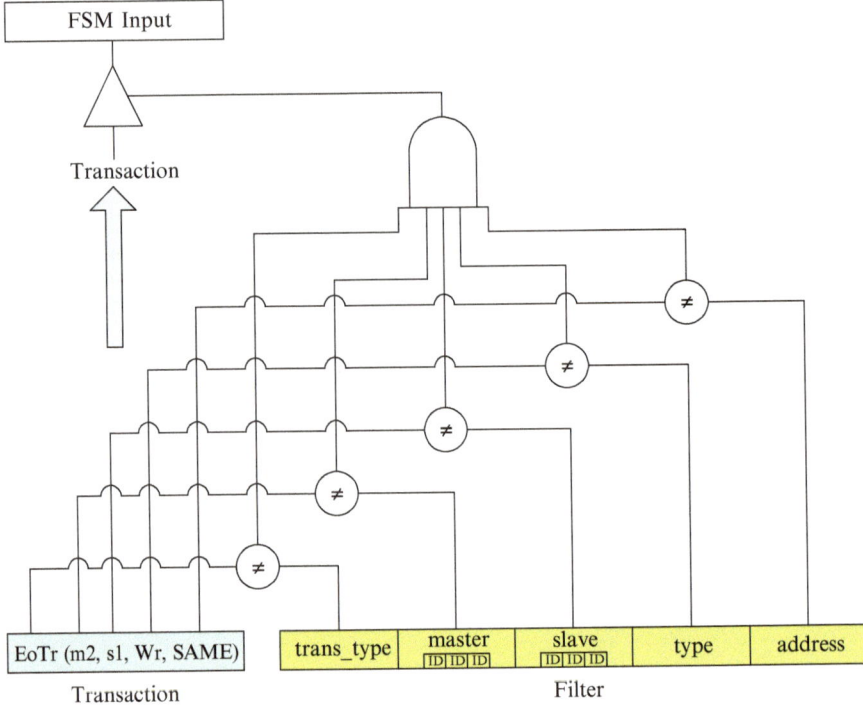

Fig. 9.6 Filter structure

9.1.5 Filter

The location of a filter is in monitors and DUs. This component is used to filter unrelated transactions in a trace. As a result, the DU receives only the related transactions for the assertion statements. Filtering can be done over all parameters of a transaction expression, i.e., *trans_type*, *master*, *slave*, *type*, and *address*. The filter is programmable according to the main assertion statements.

The filter structure is shown in Fig. 9.6. A filter has five fields: *trans_type*, *master*, *slave*, *type*, and *address*. In the fields *master* and *slave*, multiple IDs can be stored. When a transaction is received from the ordering section, the received transaction is compared to all fields of the filter. The transaction is considered as unrelated transaction if at least one element of the received transaction is equal to its corresponding field in the filter. In this case, the transaction is not passed to the FSM. Otherwise, the transaction is transferred to the FSM as a related transaction.

If a special field in the filter is not used, the corresponding field should be programmed such that its value is always guaranteed to be different from thecorresponding transaction element. Consequently, the output of the inequality operation for that field in Fig. 9.6 becomes always 1 which is a non-controlling value for the AND gate. In this case, the value of the corresponding field in the filter is called *don't care value*.

9.1.6   Debug FSM

Debug FSMs are programmable FSMs which are utilized in DUs to investigate the assertions online. Debug FSMs include local FSMs and global FSMs verifying local assertions and global assertions. DUs can be programmed to implement distributed FSMs validating online assertions in different levels. To do this, first the transaction-based assertions should be analyzed based on their locality in the corresponding SoC. Afterwards, they should be distributed among DUs (LDUs and CDU).

To implement programmable FSMs, lookup-table memory [Car01] is utilized which can be programmed in every debug round according to the new assertions.

Example 9.2. The structure of an FSM using lookup-table memory is shown in Fig. 9.7. As shown in the figure, in this method the current state bits and inputs are connected to the address bus of a lookup-table. The next state is taken from the

Address		Data
Current State	Input	Next State
Start	Tr1	A
Start	Oth	Start
A	Tr1	A
A	$Tr2_1$	B
A	Oth	Start
B	Tr1	A
B	Tr3	C
B	Oth	Start
C	Tr1	A
C	$Tr4_1$	D
C	Oth	Start
D	Tr1	A
D	$Tr5_1$	E
D	$Tr6_1$	F
D	Oth	Start
E	Tr1	A
E	$Tr6_1$	Err
E	Oth	Start
F	Tr1	A
F	$Tr5_1$	Err
F	Oth	Start
Err	-	Err

Fig. 9.7 FSM implementation using lookup-table memory

lookup-table output. The correct next states have to be stored in each location of the lookup-table to ensure the correct operation [Car01].

The total number of states in an FSM determines the number of bits for the current state. For input bits we can connect all elements of a transaction directly to the input lines of the lookup-table memory. In this case, if element *trans_type* has 1 bit (*SoTr* and *EoTr*), element *master* has 2 bits (4 IDs), element *slave* has 2 bits (4 IDs), element *type* has 1 bit (*Wr* and *rd*), and element *address* has 2 bits (SAME, SEQ, and OTHER), totally we need 8 bits for the input part of the FSM. To decrease the number of input bits, we can encode the transactions beforehand. Then we connect the encoded transaction to the input of the memory. As shown in Fig. 9.8, to encode a received transaction, we use some programmable *transaction patterns*. In each transaction pattern, all elements of one transaction are specified and stored. A received transaction is compared to the transaction patterns. If it is equal to one transaction pattern, the output of the corresponding AND gate becomes 1. Then 2^n bits input is converted into n bits output by an encoder. In Fig. 9.8, the encoder has 4 bits input and 2 bits output. The output of the encoder is connected to the input of the lookup-table memory. The goal of the encoder is to reduce the size of lookup-table memory, while the goal of the filter is to discard a group of unrelated transactions.

We use the size of the lookup-table memory, i.e., the number of address bits and data bits in Fig. 9.7, and the number of transaction patterns in Fig. 9.8 to estimate the area overhead of the assertion-based FSMs.

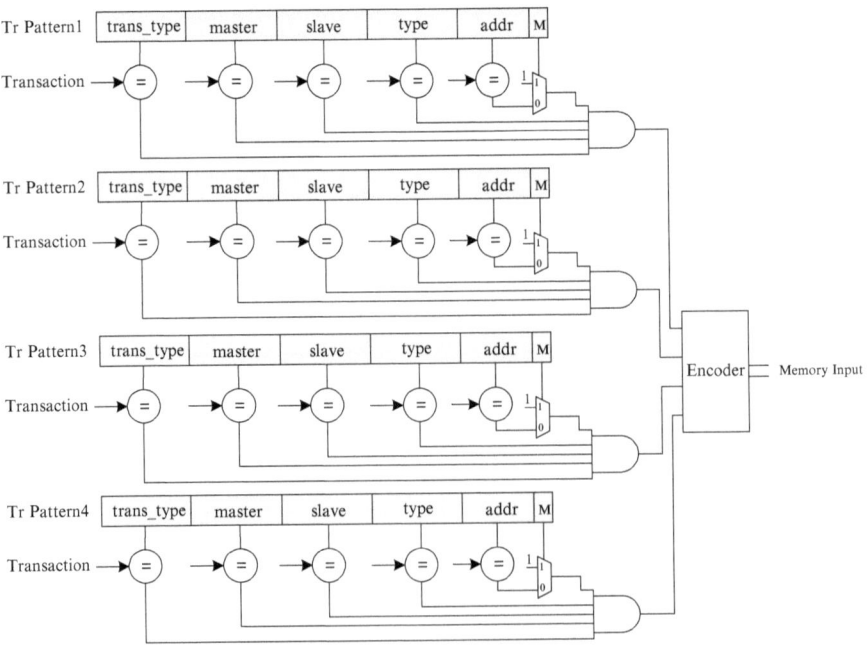

Fig. 9.8 Encoding of transaction patterns

In Fig. 9.8, field M is a mask bit for the address field. The address is considered as don't care if M is 1. Therefore, the corresponding input of the AND gate becomes 1. One mask bit can be utilized for each field of a transaction pattern. However, in Fig. 9.8 we show one mask bit only for the address field.

A complex assertion including many master and slave IDs may increase the size of the FSM. To handle this case, a complex assertion should be divided into different parts such that each part is checked by its respective LDU. The main FSM structure can be implemented in the CDU. The main FSM has some sub-FSMs. Each sub-FSM is implemented in its respective LDU. The LDUs send the state of the sub-FSMs to the CDU updating the state of the main FSM.

9.1.7 Design Decisions and Limitations

In our approach, a tree-based network is used to connect the DUs. In a tree-based network, all of the leaves have a short distance to the CDU. Therefore, the detection latency of an error is short as all transactions can arrive at the CDU in a short time. The latency of the debug network depends on the traffic of transactions and the number of hops (levels) from the leaves to the CDU. If the number of monitors is m, the tree debug network has $l = \lceil \log_2 m \rceil$ levels. In Fig. 9.2 one monitor is associated to each master. Thus, the number of monitors is $m = 4$ and the debug network has 2 levels. Using a tree-based debug network requires a large number of DUs that act like (routers when the number of masters increases. If the number of masters m is a power of two (2^n), the number of DUs in a tree-based network is $\#DU = \sum_{i=1}^{l} m/2^i = m - 1$. In each level i of the debug network, the number of DUs is $m/2^i$.

Other alternatives for the topology of the debug network can be used in order to decrease the hardware cost. One of the topologies which has been used for a debug network is ring topology [WCCC12, TGR$^+$12]. Using a ring topology for our approach decreases the number of DUs (routers) in the debug network. In the ring network, the number of required DUs is $\#DU = \lceil m/2 \rceil + 1$. We assume that each LDU can be connected to at most two monitors. The LDUs are also connected to each other in a serial manner. A CDU closes the ring. One disadvantage of the ring topology is that some LDUs have a short distance to the CDU while some other LDUs have a long distance from the CDU. The ring debug network requires dedicated techniques to balance the traffic in the network [TGR$^+$12]. In the ring network, the CDU has to wait until all of the required transactions from all the LDUs have arrived. Afterwards, the CDU can check the assertions. In the ring topology, the worst case number of hops from an LDU to the CDU is $l = \lceil \#LDUs/2 \rceil = \lceil m/4 \rceil$ which affects the latency of error detection in the CDU. Using this structure, as the CDU has to wait for the transactions of the longest path (distance), the CDU requires larger FIFOs storing all the transactions which have arrived through the shorter paths.

Another challenge is ordering the transactions of the FIFO by the CDU based on the timestamps. In [WCCC12], a nanoprocessor is used in order to investigate the transactions and their timestamps. However, in the tree network, the transactions are automatically ordered by a simple hardware mechanism from the bottom level to the top level such that at every level of the debug network, the ordered transactions are available.

The selection of the topology of the debug network is a trade-off between hardware cost and diagnosis latency. To alleviate hardware costs and to achieve a reasonable diagnosis latency, a hybrid topology may be used. A tree network is suitable for hierarchical debugging while a ring network is suitable for local debugging. The many-core systems can be divided into different clusters [WCCC12]. In each cluster, a ring subnetwork can be used to evaluate the local assertions. The local ring subnetworks are connected to each other using a tree topology evaluating the global assertions.

There are some limitations for the functionality of filters in the tree-based debug network. In the debug network, each LDU has two subnetworks, i.e., right subnetwork and left subnetwork. In this topology, a filter in an LDU can only be applied on the transactions which arrive from the masters being at the lower levels of the subnetworks of the corresponding LDU. The LDUs in the same level cannot communicate with each other. Therefore, each LDU can only check the assertions which are related to the masters falling in the leaves of the corresponding LDU. Only the CDU can have a comprehensive assertion checking the transactions of all masters.

In order to use our approach, first the address space of slaves has to be categorized in an offline step. Symbols are assigned to each relevant address range in this step. Also, in the offline step, the ID of each IP relevant for the assertion of interest is determined. The IDs are required in order to write a transaction expression.

An FSM is derived from assertions in our approach. Then, the FSMs are distributed among LDUs. Each LDU includes the FSMs which are relevant to the masters being at its lower levels. A DU can have multiple FSMs running in parallel. Parallel FSMs also help checking assertions for different address spaces. To enable this separation, each FSM can have its own filter.

9.2 Debug Flow

In our infrastructure, two kinds of assertions are used: *individual assertions* and *distributed assertions*. The individual assertions check the behavior of an individual IP by observing the corresponding interconnect. The distributed assertions are located in the DU checking the distributed behavior of the SoC against the debug patterns. Upon detection of a failure due to violating an assertion, debugging starts. As Fig. 9.9 shows, a failure can be due to violating a distributed assertion (left part of the figure) or an individual assertion (right part of the figure).

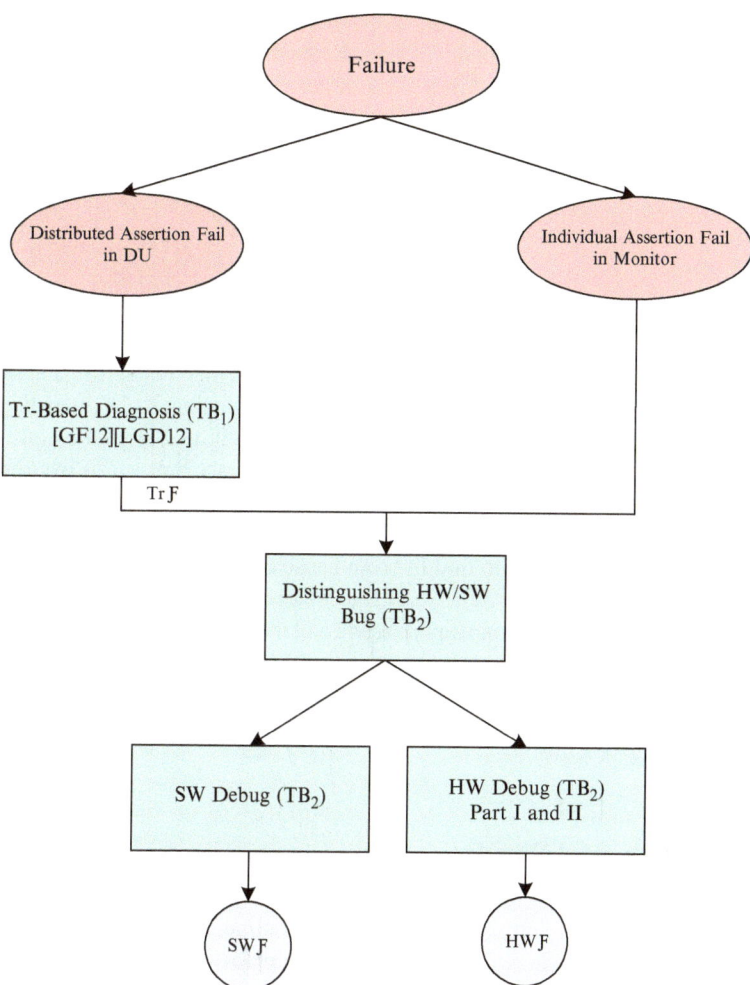

Fig. 9.9 Debug flow

We assume that there are two kinds of trace buffers in an SoC. The first kind of trace buffer is used to store the transactions at run time [GF12]. We denote this kind of trace buffer by TB_1. The trace buffer TB_1 is located on the master interconnect to store the observed transactions. Also this trace buffer can be a part of the monitor. Another kind of trace buffer is used in each IP core to store the internal traces of the corresponding IP as explained in Sect. 4.1. This kind of trace buffer is denoted by TB_2.

When a distributed assertion fails, the content of the trace buffer TB_1 is utilized to find the root cause of the failure. The trace buffer TB_1 includes the transaction traces. These traces are used to backtrace the transactions along their execution paths using BMC [GF12]. The work in [GF12] finds some explanations for the erroneous behavior by backtracing in transaction-level states. Transaction traces can also be utilized by MBD to find *Transaction Fault Candidates* ($Tr \, \mathscr{F}$). The work in [LGD12] uses MBD and an erroneous transaction trace in order to find the transaction fault candidates. They first introduce a fault model consisting of a set of rules. If a transformation of a part of the TLM fixes the error according to the defined rules, the corresponding part is considered as a transaction fault candidate.

The output of the works in [LGD12] and [GF12] is a set of transaction fault candidates. A transaction fault candidate is an erroneous transaction, function or event. However, at the transaction level the entity of a fault candidate as a hardware or software part is not specified. To enable a more accurate debugging, we need to specify whether the corresponding transaction fault candidate belongs to a hardware part or to a software part of the IP. In Fig. 9.9, after extracting $Tr \, \mathscr{F}$, we are going to distinguish hardware and software fault candidates. In this step, the content of the trace buffer TB_2 can be utilized. The content of the trace buffer TB_2 explains the behavior of an individual IP regarding the hardware and the software parts. If the content of TB_2 explains a hardware error, our hardware debug approaches explained in Parts I and II are exploited to find the root cause of the failure and to return a set of hardware fault candidates ($HW \, \mathscr{F}$). If the content of the trace buffer TB_2 implies a software error, a software debug approach is called to return a set of software fault candidates ($SW \, \mathscr{F}$).

In case an individual assertion fails, the content of the trace buffer TB_2 is utilized to distinguish hardware and software bugs and consequently to find software and hardware fault candidates.

9.3 Experimental Results

A 3×3 mesh network is setup in the Nirgam NoC simulator [Jai07] for the experiments. Nirgam is a cycle-accurate simulator which is implemented in SystemC language. We have simulated the system for one million cycles. During the run time of the SoC, our debug infrastructure continuously monitors the debug FSMs which are mentioned in the next sections. Dining philosophers [GF09] and a random application [GF12] have been implemented as example applications. In the random application, each master waits for a random time. Then the master selects a random list of slaves as resources. If the master can lock all the required resources, the processing is started. Afterwards, the resources are released and the procedure is repeated. If the master cannot lock all the required resources, the master waits for a random time and tries again [GF12].

The SoC has four masters (philosophers) and four slaves (chopsticks) in our experimental setup. The SoC IPs are divided into two groups communicating in parallel. Each group has two masters and two slaves (first group: $m1$, $s1$, $m2$, $s2$; second group: $m3$, $s3$, $m4$, $s4$). Four monitors are used to observe the master interconnects. Also two LDUs and one CDU constitute the debug network. In the experimental results, we also discuss the hardware and storage costs in terms of the size of lookup-table memory and the number of required transaction patterns. In the slave-based approach in which the address symbol is transferred by each slave, the bandwidth of the debug network is 8 *bits/cycle*, as a transaction has 8 bits: 1 bit for element *trans_type* (*SoTr* and *EoTr*), 2 bits for element *master* (4 IDs), 2 bits for element *slave* (4 IDs), 1 bit for element *type* (*Wr* and *Rd*) and 2 bits for element *address* (SAME, SEQ and OTHER). In the DU-based approach, the debug network requires a higher bandwidth, as the slave address is transferred by the debug network (Sect. 9.1.2). In total, the bandwidth of the debug network has to be $(6 + \#Address\_Bits)/cycle$.

In the following sections, we discuss debug patterns for race, deadlock, and livelock as an example. In each section first we shortly explain each pattern in TDPSL from [GF09]. Then we use the TDPSL patterns to build an FSM. Also we improve the FSMs to increase the verification coverage of the corresponding patterns. Each pattern is considered as a transaction-based assertion which is used to verify the SoC behavior online. In the following examples, we discuss the case in which the filter and the FSM only in the CDU are utilized.

9.3.1 Debug Pattern for Race

9.3.1.1 TDPSL

A race may occur in two cases [GF09]. First, one write transaction to the same place occurs during the previous write. Second, one write transaction to the same place occurs just after the end of the previous write. In TDPSL these two cases are written as following:

> *assert never*{
> $SoTr(m1, s1, Wr, -); SoTr(m2, s1, Wr, SAME); EoTr(m1, s1, Wr, SAME)$
> | $EoTr(m1, s1, Wr, -); SoTr(m2, s1, Wr, SAME)$
> }*filter*(∗, ∗, ∗)

Filtering is done on the three first parameters of transaction expressions. Sign ∗ in the filter means only the related transaction types, masters, and slaves should be considered. Therefore the transactions related to slave *s2* are omitted. Also all transactions related to the second group, i.e., transactions including master *m3* and master *m4*, have to be omitted. In our infrastructure, the filters are programmed such that the transactions related to slave *s2*, master *m3* and master *m4* are filtered online. The field *slave* in the filter of Fig. 9.6 has to include the slave ID 2. The field *master* has to include the master ID 3 and 4. Other fields in the filter have to have don't care values. The filter for the race pattern is shown in Fig. 9.10.

trans_type	master		slave	type	address
X	3	4	2	X	X

Fig. 9.10 Filter for race pattern

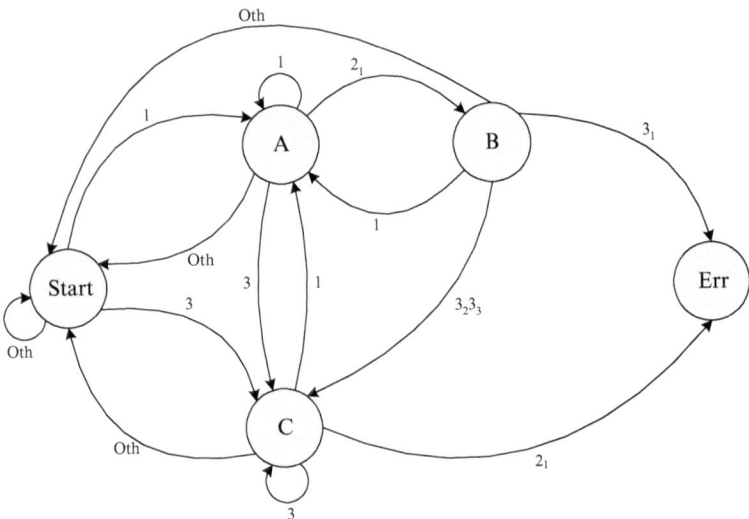

Fig. 9.11 Race FSM

As explained in Sect. 9.1.2, DRI can be transferred using a slave-based approach or a DU-based approach. In the race assertion we need the element *address* in *SoTr*. Therefore, we can apply the DU-based approach to investigate this assertion. In the salve-based approach, we the element *address* is only available when reaching the end of the transaction *EoTr*. If we use slave-based approach, we need to change the race assertion such that only *EoTr*s include element *address*. But this case causes some latency in the detection of the violation of an assertion.

9.3.1.2 FSM

To build an FSM, first we sign each transaction with a unique number. For example, transaction $SoTr(m1, s1, Wr, -)$ is written as $T1$. The address field in each transaction can be $-$, SAME, SEQ, or OTHER. We use an index for each transaction in order to distinguish different cases in field *address*. We use index 1, 2, and 3 for SAME, SEQ, and OTHER, respectively. For example $SoTr(m1, s1, Wr, SAME)$ is written as $T1_1$. Transactions $T1_1$, $T1_2$, and $T1_3$ are a subclass of transaction $T1$. By this, the race assertion is written as $T1\ T2_1\ T3_1\ |\ T3\ T2_1$. The FSM for this assertion is shown in Fig. 9.11. In the figure, only the numbers of transactions are shown. The initial state of FSM is called state *Start*. In this state, the FSM waits for transaction $T1$ or $T3$. If there is an event with transaction $T1$, the FSM switches to state A. Continuing from state A, the FSM will verify the transaction sequence $T1\ T2_1\ T3_1$. If this transaction sequence happens, the FSM goes to the error state. If in state *Start*, an event with transaction $T3$ happens, the FSM switches to state C in order to verify the sequence $T3\ T2_1$.

Table 9.1 Area overhead

Debug Pattern	Lookup Table Size (#Bits)		#Tr Patterns
	Address	Data	
Race	6	3	5
Improved Race	7	3	12
Deadlock	6	3	6
Improved Deadlock	7	4	6
Livelock	7	4	6
Improved Livelock	7	4	6

In the race FSM of Fig. 9.11, there are five states which need 3 bits from the address line of the lookup-table. Also there are five transactions $T1, T2_1, T3_1, T3_2$, and $T3_3$ ($T3$ in Fig. 9.11 is replaced by $\{T3_1, T3_2, T3_3\}$). To program a transaction pattern in Fig. 9.8 by $T1$, the mask bit of the corresponding transaction pattern is set to consider the address field as don't care.

By using the encoding structure of Fig. 9.8, five transactions are encoded into 3 bits as input bits of the lookup-table. Totally the race assertion needs 6 bits (3 bits current states + 3 bits input) for the lookup-table address and 3 bits for the lookup-table data. Also 5 transaction patterns are required (Table 9.1).

9.3.1.3 Improved FSM

To increase the verification coverage of the FSM, we need to have a more comprehensive pattern. In the following, we write an improved race pattern in TDPSL to cover more race conditions happening on slave $s1$:

assert never{
    $SoTr(m1, s1, Wr, -); SoTr(m2, s1, Wr, SAME); EoTr(m1, s1, Wr, SAME)$
    $|SoTr(m2, s1, Wr, -); SoTr(m1, s1, Wr, SAME); EoTr(m2, s1, Wr, SAME)$
    $|EoTr(m1, s1, Wr, -); SoTr(m2, s1, Wr, SAME)$
    $|EoTr(m2, s1, Wr, -); SoTr(m1, s1, Wr, SAME)$
    }*filter*$(*, *, *)$

In the first and the third line of the assertion, the pattern verifies a race condition in which master $m1$ starts a race. The second and the fourth lines of the assertion specify a race condition in which master $m2$ starts a race. We abstract this assertion as shown in following:

$$T1 \, T2_1 \, T3_1 \, |T2 \, T1_1 \, T4_1 \, | \, T3 \, T2_1 | \, T4 \, T1_1$$

Figure 9.12 shows the improved race FSM for the mentioned pattern. Eight states are used to verify this assertion online. The sequence $T1 \, T2_1 \, T3_1$ is checked by states $(Start, A1, B1, Err)$. States $(Start, A2, B2, Err)$ verify the sequence $T2 \, T1_1 \, T4_1$. The sequences $T3 \, T2_1$ and $T4 \, T1_1$ are checked by the state $(Start, C2, Err)$ and by the state $(Start, C1, Err)$, respectively.

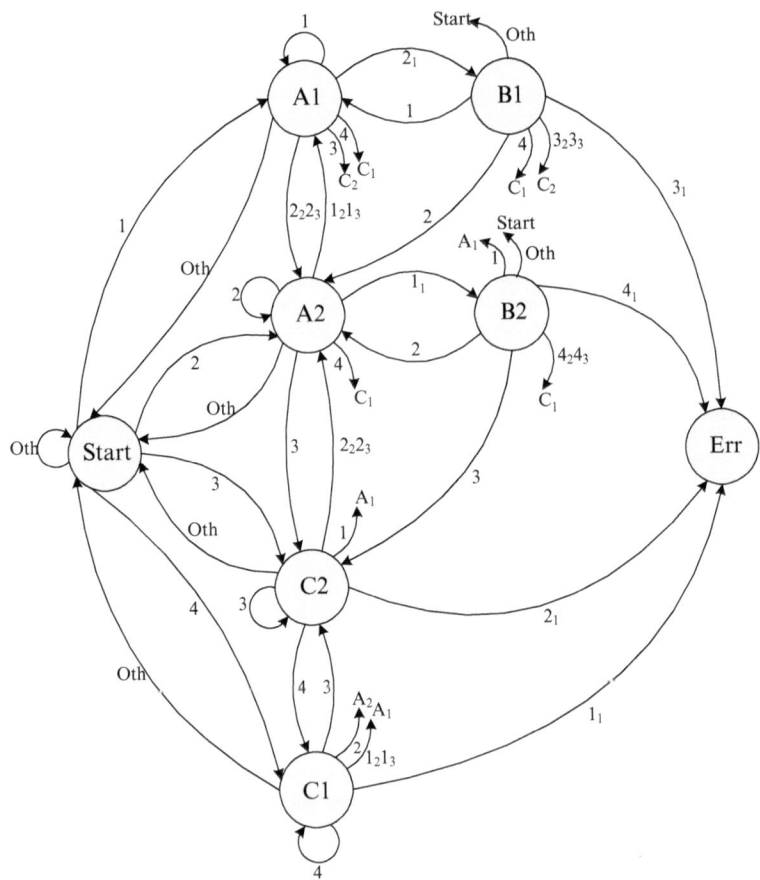

Fig. 9.12 Improved race FSM

For the improved race assertion, we need 3 bits for the FSM states. We need the following 12 transaction patterns: $T1_1$, $T1_2$, $T1_3$, $T2_1$, $T2_2$, $T2_3$, $T3_1$, $T3_2$, $T3_3$, $T4_1$, $T4_2$, and $T4_3$. Twelve transaction patterns are encoded into 4 bits. Therefore the size of the lookup-table address is 7 bits (3 + 4).

9.3.2 Debug Pattern for Deadlock

9.3.2.1 TDPSL

A deadlock happens when some masters are waiting for other masters to release shared resources. Here we show the case of two masters and two slaves as an example. Each slave has a semaphore which specifies its access permission. The slave is free when its semaphore is 0. When the semaphore is 1, the slave is locked. Each master should first lock its required slaves, then it can start its process using the corresponding slaves as resources. To lock a slave, a master has to first read the semaphore of the corresponding slave. If the semaphore is 0, then the master can write 1 to the semaphore to lock the corresponding slave. Therefore to lock a slave, a master requires two transactions, i.e., one read transaction and one write transaction. If the semaphore is 1, i.e., the slave is already locked, then the master should wait until the corresponding slave becomes released. Both masters have access to the semaphore of each slave. Accessing a semaphore is equivalent to accessing the same address by different masters.

A simple deadlock scenario for two masters and two slaves is as follows [GF09]:

1. Master1 locks the first semaphore.
2. Master2 locks the second semaphore.
3. Master1 waits for the second semaphore.
4. Master2 waits for the first semaphore.
5. Steps 3 and 4 are repeated.

This deadlock condition is written in TDPSL as follows:

$assert\ never\{$
  $EoTr(m1, s1, Rd, -); EoTr(m1, s1, Wr, SAME);$
  $EoTr(m2, s2, Rd, -); EoTr(m2, s2, Wr, SAME);$
  $\{EoTr(m1, s2, Rd, SAME); EoTr(m2, s1, Rd, SAME)$
  $| EoTr(m2, s1, Rd, SAME); EoTr(m1, s2, Rd, SAME)$
  $\}[+]$
$\}filter(*, *, *)$

In this assertion, the applications are considered in which each master first locks the slave with the same ID. For example master $m1$ first locks slave $s1$. If it is successful, then it locks slave $s2$. To implement this assertion by our debug infrastructure, the filters are programmed such that transactions $SoTr$ are filtered online as unrelated transactions. Also all transactions related to the second group, i.e., transactions related to master $m3$ and master $m4$, are filtered. In the filter structure of Fig. 9.6, the field $trans\_type$ is programmed to have $SoTr$. The field $master$ has to include the master ID 3 and 4. Other fields are programmed to have don't care values. The filter for the deadlock pattern is shown in Fig. 9.13.

trans_type	master		slave	type	address
SoTr	3	4	X	X	X

Fig. 9.13 Filter for deadlock pattern

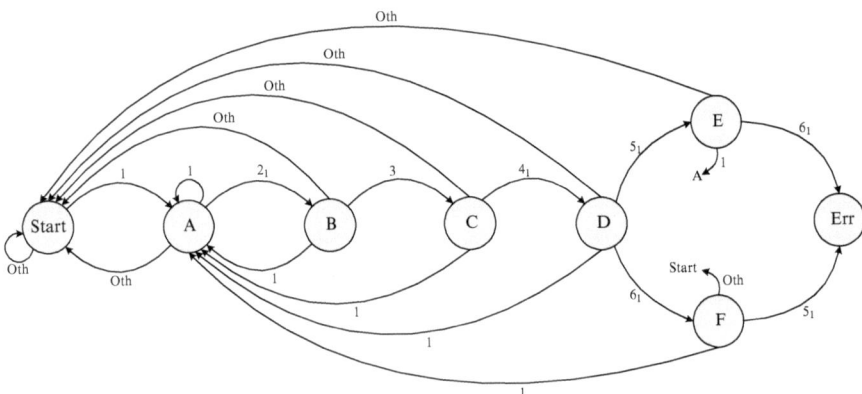

Fig. 9.14 Deadlock FSM

9.3.2.2 FSM

The deadline assertion is abstracted to $T1\,T2_1\,T3\,T4_1\{T5_1\,T6_1\,|\,T6_1\,T5_1\}$. The FSM related to the deadlock assertion is shown in Fig. 9.14. In the deadlock assertion, there is a repetition operator $+$ which means the transaction sequence between the corresponding two brackets $\{\ldots\}$ may be repeated one or more times. For the sake of simplicity, we show the case of one repetition in the FSM. The operator $+$ is implemented using a counter in the FSM which checks how many times a special transaction sequence is repeated. In Fig. 9.14, states $(Start, A, B, C, D)$ checks the transaction sequence $T1\,T2_1\,T3\,T4_1$. The transaction sequences $T5_1\,T6_1$ and $T6_1\,T5_1$ are verified by states (D, E, Err) and (D, F, Err), respectively. The FSM works correctly in the presence of suitable filters. In this case, only the related transactions are investigated by the FSM.

The deadlock FSM requires 3 bits to implement eight states. Also, it requires 6 transaction patterns: $T1$, $T2_1$, $T3$, $T4_1$, $T5_1$, and $T6_1$. The transaction patterns are encoded into 3 bits. Therefore the address line has 6 bits $(3 + 3)$. The size of lookup-table data is 3 bits.

9.3.2.3 Improved FSM

In the mentioned deadlock assertion, first the lock process from master $m1$ is checked, then the lock process from master $m2$. To illustrate this case better, we denote a read transaction (write transaction) from master m_x to slave s_x as Rxy (Wxy). In the previous deadlock assertion only the sequence $(R11, W11, R22, W22)$ is checked for the lock process. To increase the verification coverage of the deadlock assertion we check the following sequences for the lock process:

$(R11,\ W11,\ R22,\ W22)$
$(R11,\ R22,\ W11,\ W22)$

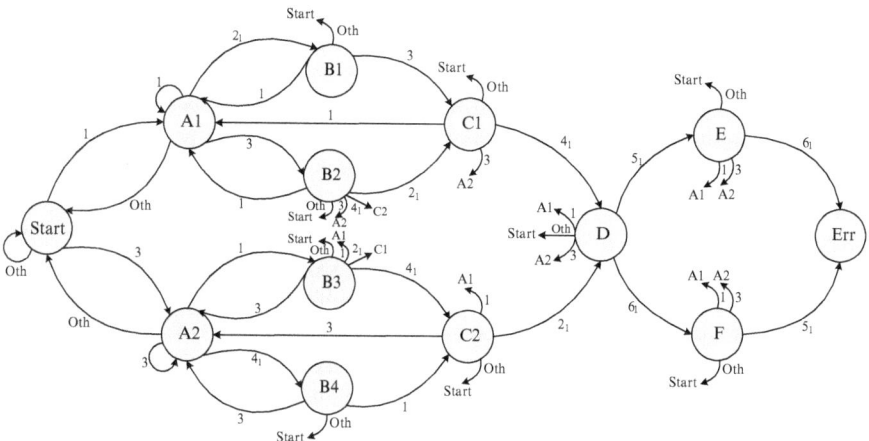

Fig. 9.15 Improved deadlock FSM

$(R11, R22, W22, W11)$
$(R22, W22, R11, W11)$
$(R22, R11, W22, W11)$
$(R22, R11, W11, W22)$

Figure 9.15 shows the improved FSM for this deadlock specification. In the improved FSM, the lock transaction sequences are verified by the following states:

$T1\ T2_1\ T3\ T4_1 \Rightarrow (Start, A1, B1, C1, D)$
$T1\ T3\ T2_1\ T4_1 \Rightarrow (Start, A1, B2, C1, D)$
$T1\ T3\ T4_1\ T2_1 \Rightarrow (Start, A1, B2, C2, D)$
$T3\ T4_1\ T1\ T2_1 \Rightarrow (Start, A2, B4, C2, D)$
$T3\ T1\ T4_1\ T2_1 \Rightarrow (Start, A2, B3, C2, D)$
$T3\ T1\ T2_1\ T4_1 \Rightarrow (Start, A2, B3, C1, D)$

The improved deadlock assertion requires 4 bits to implement 13 states. Also it uses 6 transactions (3 bits input). Thus the size of address and data are 7 bits $(4 + 3)$ and 4 bits, (Table 9.1).

9.3.3 Debug Pattern for Livelock

A livelock is similar to a deadlock which two or more processes proceed accessing shared resources which are already locked. But in the case of a livelock, they release the locked resources permitting the other processes to continue. A simple livelock scenario for two masters and two slaves is as follows [GF09]:

1. Master1 locks the first semaphore.
2. Master2 locks the second semaphore.
3. Master1 waits for the second semaphore.

4. Master2 waits for the first semaphore.
5. Master1 unlocks the first semaphore.
6. Master2 unlocks the second semaphore.
7. Steps 1 to 6 are repeated.

To implement the livelock FSM, the deadlock FSM of Fig. 9.14 can be changed to check the steps 5 and 6. The steps 5 and 6 are monitored by the following transactions:

$$EoTr(m1, s1, Wr, SAME); EoTr(m2, s2, Wr, SAME), \text{ i.e., } (W11, W22).$$

We can check the transaction sequence $(W11, W22)$ by adding two additional states to the end of the deadlock FSM. In summary, we have 10 states $(8 + 2)$. The area overhead of the livelock assertion is shown in Table 9.1.

Also we can have an improved livelock assertion by manipulating the improved deadlock FSM of Fig. 9.15. To achieve the improvement, we can add three states at the end of the improved deadlock FSM of Fig. 9.15 to check the sequences $(W11, W22)$ and $(W22, W11)$.

9.3.4   Overhead and Simulation Results

Our debug infrastructure is programmed according to race, deadlock, and livelock debug patterns and detects the occurrence of each debug pattern at run time. Table 9.1 summarizes the estimate in area overhead resulting from each of the FSMs discussed in the previous subsections.

Another aspect is the activity of the assertions. Table 9.2 presents how often each debug pattern is detected during the simulation time of one million cycles. For the random application, the approach detects the race pattern 62 times. Also at run time, this information is sent to the corresponding masters (Fig. 9.4). For the application of dining philosophers, the deadlock pattern is detected one time. In this case, after the first deadlock detection, the group of deadlocked masters cannot proceed with their process anymore.

The masters can begin a recovery process after the CDU has sent them the error state as explained in Sect. 9.1.3 (Fig. 9.5). The effect of using the recovery process in the masters is presented in Table 9.3. Without recovery process in the application of dining philosophers, the masters in one group can eat only 6 times. After that

Table 9.2   Number of debug patterns detected for each application

	Rand. Application	Din. Application
Improved Race Pattern	62	0
Improved Deadlock Pattern	0	1
Improved Livelock Pattern	0	0

Table 9.3 Effect of online recovery

	Without Recovery	With Recovery
#Eating	6	3276
#Resolved Deadlock	0	77

a deadlock happens and the masters cannot pick up their required chopsticks. However, the recovery process of Fig. 9.5 causes that the masters can continue their main process even if they get into a deadlock. In this case, a deadlock happens 77 times. Each time the CDU detects the deadlock at run time and triggers the recovery process in the masters to recover the deadlocked network. Consequently, the masters can eat more often (3276 times) as shown in Table 9.3.

The recovery example of Fig. 9.5 is suitable for software deadlocks. In the case of a hardware fault in the network which leads to a wrong routing and consequently results in a deadlock, the recovery algorithm has to select a new route avoiding the faulty route. In this case, the recovery algorithm can be enhanced using a diagnosis approach such as [GPS+12] to localize faulty components in the network.

9.4 Summary

In this chapter, an approach was introduced to online debug for NoC-based multiprocessor SoCs. Our approach includes a hardware infrastructure, debug redundant information, and FSMs. Monitors, filters, and debug units are considered in our debug hardware infrastructure. This infrastructure allows us to investigate and to debug the behavior of an NoC-based SoC at run time. Filters and FSMs are programmed according to the transaction-based assertions defined by TDPSL. In the experimental results, we investigated the performance and the efficiency of our approach for the debug patterns race, deadlock, and livelock.

Our debug infrastructure can be used for debug automation at transaction level. As our debug approach uses programmable FSMs and filters, it can be connected to a debug platform and can be programmed using the corresponding debug platform. In each round of debugging, the FSMs and filters are programmed according to the considered assertions. Then, the assertions can be changed and can be adapted according to the observed behavior of the SoC. The debug automation is achieved by integration of our hardware infrastructure with a suitable program on the debug platform.

Also using the introduced debug infrastructure, we proposed a debug flow which takes the debug approaches of Parts I and II into account. In this case, the approaches of Parts I and II can be applied at the appropriate debug step to identify the root causes of a failure to be design bugs or electrical faults.

Chapter 10
Summary and Outlook

The cost of VLSI systems verification and debugging has significantly grown in the recent years as design size and complexity have increased. Also due to time-to-market constraints, 100 % verification coverage at the design level is an elusive task. Consequently, automated debugging approaches are required at both pre-silicon and post-silicon stages in order to reduce the development time of IC products.

To automate debugging, different approaches have been proposed which are based on simulation, BDD and SAT. As SAT-based debugging has been shown as a strong approach to automate debugging, we enhanced SAT-based debugging to achieve a higher diagnosis accuracy. Also we enhanced SAT-based debugging according to the new bug and fault models appearing in both pre-silicon and post-silicon stages of the IC design.

In this book, we proposed automated debugging approaches for the bugs and the faults which appear in different abstraction levels of a hardware system, i.e., transaction-level, RTL and gate-level. In this case, our automated debug approaches are applied to a hardware system at different granularities to find the possible location of bugs and faults. Our transaction-based debug approach is applied to a hardware system at transaction-level asserting the correct relation of transactions. Our automated debug approach for design bugs finds the potential fault candidates at RTL and gate-level of a circuit. In this case, logic bugs and synchronization bugs are considered as they are the most difficult bugs to be localized. Our proposed debug automation for electrical faults (delay faults) finds the potential failing speedpaths in a circuit at gate-level.

Design bugs may be detected at both pre-silicon stage and post-silicon stage because a design bug may be missed during design verification and may slip into the chip. Electrical faults due to physical effects appear only at post-silicon stage. We proposed a unified framework to debug automation which is used at both pre-silicon and post-silicon stages. At post-silicon stage, our approach distinguishes electrical faults and design bugs. Then the approach proceeds with a suitable method to localize design bugs and electrical faults.

© Springer International Publishing Switzerland 2015
M. Dehbashi, G. Fey, *Debug Automation from Pre-Silicon to Post-Silicon*,
DOI 10.1007/978-3-319-09309-3_10

For design bugs, first we developed new methods to automate debugging of logic bugs at the pre-silicon stage. To do this, we proposed an approach for automating the design debugging procedures by integrating SAT-based debugging with testbench-based verification. The diagnosis accuracy increases by iterating debugging and counterexample generation, i.e., the total number of fault candidates decreases. The approach uses diagnostic traces to obtain more effective counterexamples and to increase the diagnosis accuracy. Then we automated the post-silicon debugging of logic bugs by integrating post-silicon trace analysis, SAT-based debugging and diagnostic trace generation. For post-silicon debugging, we use trace buffers as a hardware structure for debugging. Afterwards, we proposed a generalized framework to automate debugging at both pre-silicon and post-silicon stages.

In the pre-silicon stage, our framework also finds the potential fault candidates for synchronization bugs and differentiates them from candidates for logic bugs. For this purpose, we modeled synchronization bugs at the pre-silicon stage utilizing SAT. We proposed a suitable correction block according to the synchronization bug model. The new correction block is added to each component in the circuit. Then the approach automatically investigates the number of cycles that a signal needs to be latched earlier or later fixing the erroneous behavior of the circuit.

In our framework for post-silicon debugging of timing faults, first we modeled timing variations using SAT in the functional domain. Then our automated debugging approach diagnosed failing speedpaths under timing variations. Also we integrated static timing analysis with functional analysis to create a compact debug instance and to decrease the debugging time.

At transaction-level, we introduced a debugging infrastructure to transaction-based online debug of NoC-based SoCs without interfering in the internal components of the corresponding NoC. Our hardware infrastructure contains monitors, filters, and a debug network including DUs. Transaction-based debug patterns are found at-speed using DUs including programmable filters and FSMs. Also we proposed a debug flow which utilizes the debug infrastructure of Part III and integrates the debug approaches of Parts I and II in the flow. However, using the proposed flow renders new research directions. After detection of a failure by transaction-based assertions, we need to distinguish hardware bug and software bug because the root cause of the failure is not transparent at the transaction level. In the next step, we can utilize the hardware debug approaches of Parts I and II to localize a hardware bug. In the case of a software bug, software debug approaches can be utilized to localize a bug in the software.

The proposed debug approaches for transactions, design bugs and electrical faults have been evaluated on suitable benchmarks at different levels of abstraction, i.e., transaction-level, RTL and gate-level. The experiments have shown that our debug approaches achieve a high diagnosis accuracy and reduce the debugging time. As a result, the time of the IC development cycle decreases and the productivity of IC design increases.

References

[ABD⁺06] Miron Abramovici, Paul Bradley, Kumar Dwarakanath, Peter Levin, Gerard Memmi, and Dave Miller. A reconfigurable design-for-debug infrastructure for SoCs. In *Proceedings of the Design Automation Conference*, pages 7–12, 2006.

[ABZ03] Aseem Agarwal, David Blaauw, and Vladimir Zolotov. Statistical timing analysis for intra-die process variations with spatial correlations. In *Proceedings of the International Conference on Computer-Aided Design*, pages 900–907, 2003.

[Ait97] Robert C Aitken. Modeling the unmodelable: Algorithmic fault diagnosis. *IEEE Design & Test of Computers*, 14(3):98–103, 1997.

[AKSN07] Armin Alaghi, Naghmeh Karimi, Mahshid Sedghi, and Zainalabedin Navabi. Online NoC switch fault detection and diagnosis using a high level fault mode. In *Proceedings of the IEEE International Symposium on Defect and Fault Tolerance in VLSI Systems*, pages 21–30, 2007.

[AN07] Ehab Anis and Nicola Nicolici. Low cost debug architecture using lossy compression for silicon debug. In *Proceedings of Design, Automation and Test in Europe*, pages 1–6, 2007.

[APP10] Massimo Alioto, Gaetano Palumbo, and Melita Pennisi. Understanding the effect of process variations on the delay of static and domino logic. *IEEE Transactions on VLSI Systems*, 18(5):697–710, 2010.

[ASV⁺05] Moayad Fahim Ali, Sean Safarpour, Andreas Veneris, Magdy S Abadir, and Rolf Drechsler. Post-verification debugging of hierarchical designs. In *Proceedings of the International Conference on Computer-Aided Design*, pages 871–876, 2005.

[BBK89] Franc Brglez, David Bryan, and Krzysztof Kozminski. Combinational profiles of sequential benchmark circuits. In *Proceedings of the IEEE International Symposium on Circuits and Systems*, pages 1929–1934, 1989.

[BCSS08] David Blaauw, Kaviraj Chopra, Ashish Srivastava, and Louis Scheffer. Statistical timing analysis: From basic principles to state of the art. *IEEE Transactions on Computer-Aided Design of Integrated Circuits and Systems*, 27(4):589–607, 2008.

[BHvMW09] Armin Biere, Marijn J. H. Heule, Hans van Maaren, and Toby Walsh, editors. *Handbook of Satisfiability*, volume 185 of *Frontiers in Artificial Intelligence and Applications*. IOS Press, February 2009.

[BKWC08] Pouria Bastani, Kip Killpack, Li-C. Wang, and Eli Chiprout. Speedpath prediction based on learning from a small set of examples. In *Proceedings of the Design Automation Conference*, pages 217–222, 2008.

[BM02] Luca Benini and Giovanni De Micheli. Networks on chips: A new SoC paradigm. *IEEE Computer*, 35(1):70–78, 2002.

© Springer International Publishing Switzerland 2015
M. Dehbashi, G. Fey, *Debug Automation from Pre-Silicon to Post-Silicon*,
DOI 10.1007/978-3-319-09309-3

[BPH85] Franc Brglez, Phillip Pownall, and Robert Hum. Accelerated ATPG and fault grading via testability analysis. In *Proceedings of the IEEE International Symposium on Circuits and Systems*, pages 695–698, 1985.

[Bre10] Melvin A. Breuer. Hardware that produces bounded rather than exact results. In *Proceedings of the Design Automation Conference*, pages 871–876, 2010.

[Brg85] Franc Brglez. A fast fault grader: Analysis and applications. In *Proceedings of the International Test Conference*, pages 785–794, 1985.

[CAK+09] Caroline Concatto, Pedro Almeida, Fernanda Lima Kastensmidt, Érika F. Cota, Marcelo Lubaszewski, and Marcos Herve. Improving yield of torus NoCs through fault-diagnosis-and-repair of interconnect faults. In *IEEE International On-Line Testing Symposium*, pages 61–66, 2009.

[Car01] John D Carpinelli. *Computer systems organization & architecture*. Addison-Wesley Boston, San Francisco, New York, 2001.

[CCLL11] Yung-Chang Chang, Ching-Te Chiu, Shih-Yin Lin, and Chung-Kai Liu. On the design and analysis of fault tolerant NoC architecture using spare routers. In *Proceedings of the ASP Design Automation Conference*, pages 431–436, 2011.

[CH97] Pi-Yu Chung and Ibrahim N. Hajj. Diagnosis and correction of multiple logic design errors in digital circuits. *IEEE Transactions on VLSI Systems*, 5(2):233–237, 1997.

[CKY03] Edmund M. Clarke, Daniel Kroening, and Karen Yorav. Specifying and verifying systems with multiple clocks. In *International Conference on Computer Design*, pages 48–55, 2003.

[CMA08] Kypros Constantinides, Onur Mutlu, and Todd M. Austin. Online design bug detection: RTL analysis, flexible mechanisms, and evaluation. In *International Symposium on Microarchitecture*, pages 282–293, 2008.

[CMAB09] Kypros Constantinides, Onur Mutlu, Todd M. Austin, and Valeria Bertacco. A flexible software-based framework for online detection of hardware defects. *IEEE Transactions Computers*, 58(8):1063–1079, 2009.

[CMB07a] Kai-Hui Chang, Igor L. Markov, and Valeria Bertacco. Automating post-silicon debugging and repair. In *Proceedings of the International Conference on Computer-Aided Design*, pages 91–98, 2007.

[CMB07b] Kai-Hui Chang, Igor L Markov, and Valeria Bertacco. Fixing design errors with counterexamples and resynthesis. In *Proceedings of the ASP Design Automation Conference*, pages 944–949, 2007.

[CMR+10] Vinay K. Chippa, Debabrata Mohapatra, Anand Raghunathan, Kaushik Roy, and Srimat T. Chakradhar. Scalable effort hardware design: exploiting algorithmic resilience for energy efficiency. In *Proceedings of the Design Automation Conference*, pages 555–560, 2010.

[Con14] Concept Engineering GmbH, http://www.concept.de/RTLvision.html [accessed: 2014-05-20]. *RTLvision PRO*, 2014.

[CR10] Srimat T. Chakradhar and Anand Raghunathan. Best-effort computing: re-thinking parallel software and hardware. In *Proceedings of the Design Automation Conference*, pages 865–870, 2010.

[CSMSV10] Yibin Chen, Sean Safarpour, Joao Marques-Silva, and Andreas Veneris. Automated design debugging with maximum satisfiability. *IEEE Transactions on Computer-Aided Design of Integrated Circuits and Systems*, 29(11):1804–1817, 2010.

[Dav99] S. Davidson. *ITC99 Benchmark*. http://www.cerc.utexas.edu/itc99-benchmarks/bench.html [accessed: 2014-05-20], 1999.

[DF12] Mehdi Dehbashi and Goerschwin Fey. Automated debugging from pre-silicon to post-silicon. In *IEEE Symposium on Design and Diagnostics of Electronic Circuits and Systems*, pages 324–329, 2012.

[DF13a] Mehdi Dehbashi and Goerschwin Fey. Debug automation for logic circuits under timing variations. *IEEE Design & Test of Computers*, 30(6):60–69, 2013.

[DF13b] Mehdi Dehbashi and Goerschwin Fey. Efficient automated speedpath debugging. In *IEEE Symposium on Design and Diagnostics of Electronic Circuits and Systems*, pages 48–53, 2013.

[DF14a] Mehdi Dehbashi and Goerschwin Fey. Debug automation for synchronization bugs at RTL. In *Proceedings of the International Conference on VLSI Design*, 2014.

[DF14b] Mehdi Dehbashi and Goerschwin Fey. Transaction-based online debug for NoC-based multiprocessor SoCs. In *Euromicro Conference on Parallel, Distributed, and Network-Based Processing (PDP)*, 2014.

[DFRR12] Mehdi Dehbashi, Goerschwin Fey, Kaushik Roy, and Anand Raghunathan. On modeling and evaluation of logic circuits under timing variations. In *EUROMICRO Symp. on Digital System Design*, pages 431–436, 2012.

[dKK03] Johan de Kleer and James Kurien. Fundamentals of model-based diagnosis. In *IFAC Symposium on Fault Detection, Supervision, and Safety of Technical Processes*, pages 25–36, 2003.

[dPNN+11] Flavio M. de Paula, Amir Nahir, Ziv Nevo, Avigal Orni, and Alan J. Hu. TAB-BackSpace: Unlimited-length trace buffers with zero additional on-chip overhead. In *Proceedings of the Design Automation Conference*, pages 411–416, 2011.

[DSF11] Mehdi Dehbashi, André Sülflow, and Goerschwin Fey. Automated design debugging in a testbench-based verification environment. In *EUROMICRO Symp. on Digital System Design*, pages 479–486, 2011.

[DSF13] Mehdi Dehbashi, André Sülflow, and Goerschwin Fey. Automated design debugging in a testbench-based verification environment. *Microprocessors and Microsystems*, 37(2):206–217, 2013.

[EEH+06] Wolfgang Ecker, Volkan Esen, Michael Hull, Thomas Steininger, and Michael Velten. Requirements and concepts for transaction level assertions. In *International Conference on Computer Design*, 2006.

[EKD+03] Dan Ernst, Nam Sung Kim, Shidhartha Das, Sanjay Pant, Rajeev R. Rao, Toan Pham, Conrad H. Ziesler, David Blaauw, Todd M. Austin, Krisztián Flautner, and Trevor N. Mudge. Razor: A low-power pipeline based on circuit-level timing speculation. In *International Symposium on Microarchitecture*, pages 7–18, 2003.

[ES04] Niklas Eén and Niklas Sörensson. An extensible SAT solver. In *Proceedings of the International Conference on Theory and Applications of Satisfiability Testing*, volume 2919 of *Lecture Notes in Computer Science*, pages 502–518, 2004.

[FAVS+04] M Fahim Ali, Andreas Veneris, Alexander Smith, Sean Safarpour, Rolf Drechsler, and Magdy Abadir. Debugging sequential circuits using Boolean satisfiability. In *Proceedings of the International Conference on Computer-Aided Design*, pages 204–209, 2004.

[FD05] Goerschwin Fey and Rolf Drechsler. Efficient hierarchical system debugging for property checking. In *In IEEE Workshop on Design and Diagnostics of Electronic Circuits and Systems, 2005*, pages 41–46, 2005.

[FSBD08] Goerschwin Fey, Stefan Staber, Roderick Bloem, and Rolf Drechsler. Automatic fault localization for property checking. *IEEE Transactions on Computer-Aided Design of Integrated Circuits and Systems*, 27(6):1138–1149, 2008.

[FSW99] Gerhard Friedrich, Markus Stumptner, and Franz Wotawa. Model-based diagnosis of hardware designs. *Artificial Intelligence*, 111(1–2):3–39, 1999.

[GF09] Amir Masoud Gharehbaghi and Masahiro Fujita. Transaction-based debugging of system-on-chips with patterns. In *International Conference on Computer Design*, pages 186–192, 2009.

[GF12] Amir Masoud Gharehbaghi and Masahiro Fujita. Transaction-based post-silicon debug of many-core system-on-chips. In *Proceedings of the International Symposium on Quality Electronic Design*, pages 702–708, 2012.

[GG07] Malay K. Ganai and Aarti Gupta. Efficient BMC for multi-clock systems with clocked specifications. In *Proceedings of the ASP Design Automation Conference*, pages 310–315, 2007.

[GK05] Alex Groce and Daniel Kroening. Making the most of BMC counterexamples. *Electronic Notes in Theoretical Computer Science*, 119(2):67–81, 2005.

[GK10] Kunal P. Ganeshpure and Sandip Kundu. On ATPG for multiple aggressor crosstalk faults. *IEEE Transactions on Computer-Aided Design of Integrated Circuits and Systems*, 29(5):774–787, 2010.

[GMK91] Torsten Grüning, Udo Mahlstedt, and Hartmut Koopmeiners. DIATEST: A fast diagnostic test pattern generator for combinational circuits. In *Proceedings of the International Conference on Computer-Aided Design*, pages 194–197, 1991.

[GMP+11] Vaibhav Gupta, Debabrata Mohapatra, Sang Phill Park, Anand Raghunathan, and Kaushik Roy. IMPACT: imprecise adders for low-power approximate computing. In *International Symposium on Low Power Electronics and Design*, pages 409–414, 2011.

[GPS+12] Amirali Ghofrani, Ritesh Parikh, Saeed Shamshiri, Andrew DeOrio, Kwang-Ting Cheng, and Valeria Bertacco. Comprehensive online defect diagnosis in on-chip networks. In *Proceedings of the VLSI Test Symposium*, pages 44–49, 2012.

[GSY07] Orna Grumberg, Assaf Schuster, and Avi Yadgar. 3-valued circuit SAT for STE with automatic refinement. In *Automated Technology for Verification and Analysis*, pages 457–473, 2007.

[Gup07] Aarti Gupta. *SAT-based scalable formal verification solutions*. Springer, 2007.

[GVVSB07] Kees Goossens, Bart Vermeulen, Remco Van Steeden, and Martijn Bennebroek. Transaction-based communication-centric debug. In *International Symposium on Networks-on-Chips*, pages 95–106, 2007.

[GYP+10] Mingzhi Gao, Zuochang Ye, Yao Peng, Yan Wang, and Zhiping Yu. A comprehensive model for gate delay under process variation and different driving and loading conditions. In *Proceedings of the International Symposium on Quality Electronic Design*, pages 406–412, 2010.

[Hay85] John P. Hayes. Fault modeling. *IEEE Design & Test of Computers*, pages 37–44, 1985.

[HMM06] Andrew B. T. Hopkins and Klaus D. McDonald-Maicr. Debug support for complex systems on-chip: a review. *IEE Proceedings on Computers and Digital Techniques*, 153(4):197 – 207, 2006.

[IEE05] IEEE. *IEEE Std 1850–2005 – IEEE Standard for Property Specification Language (PSL)*. The IEEE, 2005.

[Jai07] L. Jain. *NIRGAM: A Simulator for NoC Interconnect Routing and Application Modeling – Version 1.1*, 2007. http://nirgam.ecs.soton.ac.uk/ [accessed: 2014-05-20].

[JLJ09] Tai-Ying Jiang, C-NJ Liu, and Jing-Yang Jou. Accurate rank ordering of error candidates for efficient HDL design debugging. *IEEE Transactions on Computer-Aided Design of Integrated Circuits and Systems*, 28(2):272–284, 2009.

[KKC07] Kip Killpack, Chandramouli V. Kashyap, and Eli Chiprout. Silicon speedpath measurement and feedback into EDA flows. In *Proceedings of the Design Automation Conference*, pages 390–395, 2007.

[KKKS10] Andrew B. Kahng, Seokhyeong Kang, Rakesh Kumar, and John Sartori. Slack redistribution for graceful degradation under voltage overscaling. In *Proceedings of the ASP Design Automation Conference*, pages 825–831, 2010.

[KM03] JA Knottnerus and JW Muris. Assessment of the accuracy of diagnostic tests: the cross-sectional study. *Journal of Clinical Epidemiology*, 56(11):1118–1128, 2003.

[KN08] Ho Fai Ko and Nicola Nicolici. Automated trace signals identification and state restoration for improving observability in post-silicon validation. In *Proceedings of Design, Automation and Test in Europe*, pages 1298–1303, 2008.

[KNKB08] Kip Killpack, Suriyaprakash Natarajan, Arun Krishnamachary, and Pouria Bastani. Case study on speed failure causes in a microprocessor. *IEEE Design & Test of Computers*, 25(3):224–230, 2008.

[Lar92] Tracy Larrabee. Test pattern generation using boolean satisfiability. *IEEE Transactions on Computer-Aided Design of Integrated Circuits and Systems*, 11(1): 4–15, 1992.

[LB14] Damjan Lampret and Julius Baxter. *OpenRISC 1200 IP Core Specification (Preliminary Draft)*, 2014. http://www.openrisc.net [accessed: 2014-05-20].

[LCB+10] Larkhoon Leem, Hyungmin Cho, Jason Bau, Quinn A. Jacobson, and Subhasish Mitra. ERSA: Error resilient system architecture for probabilistic applications. In *Proceedings of Design, Automation and Test in Europe*, pages 1560–1565, 2010.

[LDX12] Min Li, Azadeh Davoodi, and Lin Xie. Custom on-chip sensors for post-silicon failing path isolation in the presence of process variations. In *Proceedings of Design, Automation and Test in Europe*, pages 1591–1596, 2012.

[LEN+11] Avinash Lingamneni, Christian Enz, Jean-Luc Nagel, Krishna Palem, and Christian Piguet. Energy parsimonious circuit design through probabilistic pruning. In *Proceedings of Design, Automation and Test in Europe*, pages 764–769, 2011.

[LGD12] Hoang M. Le, Daniel Große, and Rolf Drechsler. Automatic TLM fault localization for SystemC. *IEEE Transactions on Computer-Aided Design of Integrated Circuits and Systems*, 31(8):1249–1262, 2012.

[LLC07] Yung-Chieh Lin, Feng Lu, and Kwang-Ting Cheng. Multiple-fault diagnosis based on adaptive diagnostic test pattern generation. *IEEE Transactions on Computer-Aided Design of Integrated Circuits and Systems*, 26(5):932–942, 2007.

[LMF11] Yeonbok Lee, Takeshi Matsumoto, and Masahiro Fujita. On-chip dynamic signal sequence slicing for efficient post-silicon debugging. In *Proceedings of the ASP Design Automation Conference*, pages 719–724, 2011.

[LRS89] Wing-Ning Li, Sudhakar M Reddy, and Sartaj K Sahni. On path selection in combinational logic circuits. *IEEE Transactions on Computer-Aided Design of Integrated Circuits and Systems*, 8(1):56–63, 1989.

[LS80] Jean Davies Lesser and John J. Shedletsky. An experimental delay test generator for LSI logic. *IEEE Transactions on Computers*, 100(3):235–248, 1980.

[LV05] Jiang Brandon Liu and Andreas Veneris. Incremental fault diagnosis. *IEEE Transactions on Computer-Aided Design of Integrated Circuits and Systems*, 24(2): 240–251, 2005.

[LWPM05] Leonard Lee, Li-C. Wang, Praveen Parvathala, and T. M. Mak. On silicon-based speed path identification. In *Proceedings of the VLSI Test Symposium*, pages 35–41, 2005.

[LX10] Xiao Liu and Qiang Xu. On signal tracing for debugging speedpath-related electrical errors in post-silicon validation. In *Proceedings of the IEEE Asian Test Symposium*, pages 243–248, 2010.

[Mal87] Wojciech Maly. Realistic fault modeling for VLSI testing. In *Proceedings of the Design Automation Conference*, pages 173–180. ACM, 1987.

[MB91] Patrick C McGeer and Robert K Brayton. *Integrating Functional and Temporal Domains in Logic Design: The False Path Problem and Its Implications*. Kluwer Academic Publishers, 1991.

[McE93] K. McElvain. *IWLS'93 Benchmark Set: Version 4.0.* http://www.cbl.ncsu.edu/benchmarks/LGSynth93 [accessed: 2014-05-20], 1993.

[MCRR11] Debabrata Mohapatra, Vinay K. Chippa, Anand Raghunathan, and Kaushik Roy. Design of voltage-scalable meta-functions for approximate computing. In *Proceedings of Design, Automation and Test in Europe*, pages 950–955, 2011.

[MJR86] Yashwant K Malaiya, AP Jayasumana, and R Rajsuman. A detailed examination of bridging faults. In *International Conference on Computer Design*, pages 78–81, 1986.

[MMSTR09] Vishal J. Mehta, Malgorzata Marek-Sadowska, Kun-Han Tsai, and Janusz Rajski. Timing-aware multiple-delay-fault diagnosis. *IEEE Transactions on Computer-Aided Design of Integrated Circuits and Systems*, 28(2):245–258, 2009.

[MS95] Joao Marques-Silva. *Search algorithms for satisfiability problems in combinational switching circuits*. PhD thesis, University of Michigan, 1995.

[MS07] Wolfgang Mayer and Markus Stumptner. Model-based debugging–state of the art and future challenges. *Electronic Notes in Theoretical Computer Science*, 174(4): 61–82, 2007.

[MVL09] Richard McLaughlin, Srikanth Venkataraman, and Carlston Lim. Automated debug of speed path failures using functional tests. In *Proceedings of the VLSI Test Symposium*, pages 91–96, 2009.

[MVS⁺07] Hratch Mangassarian, Andreas Veneris, Sean Safarpour, Farid N Najm, and Magdy S Abadir. Maximum circuit activity estimation using pseudo-boolean satisfiability. In *Proceedings of Design, Automation and Test in Europe*, pages 1538–1543, 2007.

[Nan11] Nangate. *Nangate 45nm Open Cell Library*, 2011. http://www.nangate.com [accessed: 2014-05-20].

[OHN09] Sari Onaissi, Khaled R. Heloue, and Farid N. Najm. PSTA-based branch and bound approach to the silicon speedpath isolation problem. In *Proceedings of the International Conference on Computer-Aided Design*, pages 217–224, 2009.

[Ope09] Open SystemC Initiative. *TLM-2.0 Language Reference Manual*, 2009. http://www.systemc.org [accessed: 2014-05-20].

[PGI⁺05] Partha Pratim Pande, Cristian Grecu, André Ivanov, Resve A. Saleh, and Giovanni De Micheli. Design, synthesis, and test of networks on chips. *IEEE Design & Test of Computers*, 22(5):404–413, 2005.

[PGSR10] Mihalis Psarakis, Dimitris Gizopoulos, Edgar Sánchez, and Matteo Sonza Reorda. Microprocessor software-based self-testing. *IEEE Design & Test of Computers*, 27(3):4–19, 2010.

[PHM09] Sung-Boem Park, Ted Hong, and Subhasish Mitra. Post-silicon bug localization in processors using instruction footprint recording and analysis (IFRA). *IEEE Transactions on Computer-Aided Design of Integrated Circuits and Systems*, 28(10): 1545–1558, 2009.

[QW03] Wangqi Qiu and D. M. H. Walker. An efficient algorithm for finding the K longest testable paths through each gate in a combinational circuit. In *Proceedings of the International Test Conference*, pages 592–601, 2003.

[Rei87] Raymond Reiter. A theory of diagnosis from first principles. *Artificial Intelligence*, 32:57–95, 1987.

[RGU09] Jaan Raik, Vineeth Govind, and Raimund Ubar. Design-for-testability-based external test and diagnosis of mesh-like network-on-a-chips. *IET Computers & Digital Techniques*, 3(5):476–486, 2009.

[RS04] Kavita Ravi and Fabio Somenzi. Minimal assignments for bounded model checking. In *Tools and Algorithms for the Construction and Analysis of Systems*, volume 2988 of *LNCS*, pages 31–45, 2004.

[SAKJ10] Naresh R. Shanbhag, Rami A. Abdallah, Rakesh Kumar, and Douglas L. Jones. Stochastic computation. In *Proceedings of the Design Automation Conference*, pages 859–864, 2010.

[SCPB12] Matthias Sauer, Alexander Czutro, Ilia Polian, and Bernd Becker. Small-delay-fault ATPG with waveform accuracy. In *Proceedings of the International Conference on Computer-Aided Design*, pages 30–36, 2012.

[SD10] André Sülflow and Rolf Drechsler. Automatic fault localization for programmable logic controllers. In *Formal Methods for Automation and Safety in Railway and Automotive Systems*, pages 247–256, 2010.

[SFB⁺09] André Sülflow, Goerschwin Fey, Cécile Braunstein, Ulrich Kühne, and Rolf Drechsler. Increasing the accuracy of SAT-based debugging. In *Proceedings of Design, Automation and Test in Europe*, pages 1326–1332, 2009.

[SFBD08] André Sülflow, Goerschwin Fey, Roderick Bloem, and Rolf Drechsler. Using unsatisfiable cores to debug multiple design errors. In *Great Lakes Symposium VLSI*, pages 77–82, 2008.

[SFD10] André Sülflow, Goerschwin Fey, and Rolf Drechsler. Using QBF to increase accuracy of SAT-based debugging. In *Proceedings of the IEEE International Symposium on Circuits and Systems*, pages 641–644, 2010.

[SG10] Doochul Shin and Sandeep K. Gupta. Approximate logic synthesis for error tolerant applications. In *Proceedings of Design, Automation and Test in Europe*, pages 957–960, 2010.

[SG11] Doochul Shin and Sandeep K. Gupta. A new circuit simplification method for error tolerant applications. In *Proceedings of Design, Automation and Test in Europe*, pages 1566–1571, 2011.

[SGC11] Saeed Shamshiri, Amirali Ghofrani, and Kwang-Ting Cheng. End-to-end error correction and online diagnosis for on-chip networks. In *Proceedings of the International Test Conference*, pages 1–10, 2011.

[SGT⁺08] Smruti R. Sarangi, Brian Greskamp, Radu Teodorescu, Jun Nakano, Abhishek Tiwari, and Josep Torrellas. VARIUS: A model of process variation and resulting timing errors for microarchitects. *IEEE Transactions Semiconductor Manufacturing*, 21(1): 3–13, 2008.

[SGTT08] Smruti R. Sarangi, Brian Greskamp, Abhishek Tiwari, and Josep Torrellas. EVAL: Utilizing processors with variation-induced timing errors. In *International Symposium on Microarchitecture*, pages 423–434, 2008.

[SKF⁺09] André Sülflow, Ulrich Kuhne, Goerschwin Fey, Daniel Grosse, and Rolf Drechsler. WoLFram – a word level framework for formal verification. In *IEEE/IFIP International Symposium on Rapid System Prototyping*, pages 11–17, 2009.

[SRL⁺11] Alessandro Strano, Crispín Gómez Requena, Daniele Ludovici, Michele Favalli, María Engracia Gómez, and Davide Bertozzi. Exploiting network-on-chip structural redundancy for a cooperative and scalable built-in self-test architecture. In *Proceedings of Design, Automation and Test in Europe*, pages 661–666, 2011.

[SV07] Sean Safarpour and Andreas Veneris. Abstraction and refinement techniques in automated design debugging. In *Proceedings of Design, Automation and Test in Europe*, pages 1182–1187, 2007.

[SVAV05] Alexander Smith, Andreas Veneris, Moayad Fahim Ali, and Anastasios Viglas. Fault diagnosis and logic debugging using Boolean satisfiability. *IEEE Transactions on Computer-Aided Design of Integrated Circuits and Systems*, 24(10):1606–1621, 2005.

[SVD08] Sean Safarpour, Andreas G Veneris, and Rolf Drechsler. Improved SAT-based reachability analysis with observability don't cares. *Journal of Satisfiability, Boolean Modeling and Computation*, 5:1–25, 2008.

[Syn14] Synopsys Inc. *Design Vision - Synopsys Inc.*, 2014. http://www.synopsys.com [accessed: 2014-05-20].

[TBG11] Desta Tadesse, R. Iris Bahar, and Joel Grodstein. Test vector generation for post-silicon delay testing using SAT-based decision problems. *Journal of Electronic Testing: Theory and Applications*, 27(2):123–136, 2011.

[TBW⁺09] James Tschanz, Keith A. Bowman, Chris Wilkerson, Shih-Lien Lu, and Tanay Karnik. Resilient circuits - enabling energy-efficient performance and reliability. In *Proceedings of the International Conference on Computer-Aided Design*, pages 71–73, 2009.

[TGR⁺12] Vladimir Todorov, Alberto Ghiribaldi, Helmut Reinig, Davide Bertozzi, and Ulf Schlichtmann. Non-intrusive trace & debug NoC architecture with accurate timestamping for GALS SoCs. In *International Conference on Hardware/Software Codesign and System Synthesis*, pages 181–186, 2012.

[THPM⁺10] Yanjing Li Ted Hong, Sung-Boem Park, Diana Mui, David Lin, Ziyad Abdel Kaleq, Nagib Hakim, Helia Naeimi, Donald S. Gardner, and Subhasish Mitra. QED: Quick error detection tests for effective post-silicon validation. In *Proceedings of the International Test Conference*, pages 1–10, 2010.

[Tse68] Grigori S Tseitin. On the complexity of derivation in the propositional calculus. *Zapiski nauchnykh seminarov LOMI*, 8:234–259, 1968.

[TX07] Shan Tang and Qiang Xu. A multi-core debug platform for NoC-based systems. In *Proceedings of Design, Automation and Test in Europe*, pages 870–875, 2007.

[VARR11] Rangharajan Venkatesan, Amit Agarwal, Kaushik Roy, and Anand Raghunathan. MACACO: Modeling and analysis of circuits for approximate computing. In *Proceedings of the International Conference on Computer-Aided Design*, pages 667–673, 2011.

[Vel05] Miroslav N Velev. Comparison of schemes for encoding unobservability in translation to SAT. In *Proceedings of the ASP Design Automation Conference*, pages 1056–1059, 2005.

[Ver02] SystemC Version. 2.0 user's guide. *Open SystemC Initiative*, 2002.

[VG09] Bart Vermeulen and Kees Goossens. A network-on-chip monitoring infrastructure for communication-centric debug of embedded multi-processor SoCs. In *International Symposium on VLSI Design, Automation and Test*, pages 183–186, 2009.

[VH99] Andreas Veneris and Ibrahim N Hajj. Design error diagnosis and correction via test vector simulation. *IEEE Transactions on Computer-Aided Design of Integrated Circuits and Systems*, 18(12):1803–1816, 1999.

[VWB02] Baart Vermeulen, Tom Waayers, and Sjaak Bakker. IEEE 1149.1-compliant access architecture for multiple core debug on digital system chips. In *Proceedings of the International Test Conference*, pages 55–63, 2002.

[VWG02] Bart Vermeulen, Tom Waayers, and Sandeep Kumar Goel. Core-based scan architecture for silicon debug. In *Proceedings of the International Test Conference*, pages 638–647, 2002.

[WC09] Lu Wan and Deming Chen. Dynatune: Circuit-level optimization for timing speculation considering dynamic path behavior. In *Proceedings of the International Conference on Computer-Aided Design*, pages 172–179, 2009.

[WCCC12] Chi-Neng Wen, Shu-Hsuan Chou, Chien-Chih Chen, and Tien-Fu Chen. NUDA: A non-uniform debugging architecture and nonintrusive race detection for many-core systems. *IEEE Transactions Computers*, 61(2):199–212, 2012.

[WLRI87] John A Waicukauski, Eric Lindbloom, Barry K Rosen, and Vijay S Iyengar. Transition fault simulation. *IEEE Design & Test of Computers*, 4(2):32–38, 1987.

[XD10] Lin Xie and Azadeh Davoodi. Representative path selection for post-silicon timing prediction under variability. In *Proceedings of the Design Automation Conference*, pages 386–391, 2010.

[XD11] Lin Xie and Azadeh Davoodi. Bound-based statistically-critical path extraction under process variations. *IEEE Transactions on Computer-Aided Design of Integrated Circuits and Systems*, 30(1):59–71, 2011.

[XDS10] Lin Xie, Azadeh Davoodi, and Kewal K. Saluja. Post-silicon diagnosis of segments of failing speedpaths due to manufacturing variations. In *Proceedings of the Design Automation Conference*, pages 274–279, 2010.

[YNV09] Yu-Shen Yang, Nicola Nicolici, and Andreas G. Veneris. Automated data analysis solutions to silicon debug. In *Proceedings of Design, Automation and Test in Europe*, pages 982–987, 2009.

[YPK10] Hyunbean Yi, Sungju Park, and Sandip Kundu. On-chip support for NoC-based SoC debugging. *IEEE Transactions on Circuits and Systems*, 57-I(7):1608–1617, 2010.

[YT08] Joon-Sung Yang and Nur A. Touba. Expanding trace buffer observation window for in-system silicon debug through selective capture. In *Proceedings of the VLSI Test Symposium*, pages 345–351, 2008.

[YT13] Joon-Sung Yang and Nur A Touba. Improved trace buffer observation via selective data capture using 2-d compaction for post-silicon debug. *Very Large Scale Integration (VLSI) Systems, IEEE Transactions*, 21(2):320–328, 2013.

[ZCY+07] Feijun Zheng, Kwang-Ting Cheng, Xiaolang Yan, John Moondanos, and Ziyad Hanna. An efficient diagnostic test pattern generation framework using boolean satisfiability. In *Proceedings of the ASP Design Automation Conference*, pages 288–294, 2007.

[ZGC+10] Jing Zeng, Ruifeng Guo, Wu-Tung Cheng, Michael Mateja, Jing Wang, Kun-Han Tsai, and Ken Amstutz. Scan based speed-path debug for a microprocessor. In *European Test Symposium*, pages 207–212, 2010.

Index

© Springer International Publishing Switzerland 2015
M. Dehbashi, G. Fey, *Debug Automation from Pre-Silicon to Post-Silicon*,
DOI 10.1007/978-3-319-09309-3